The Science of
THE BEATLES

The Science of
THE BEATLES

THE TECHNOLOGY AND THEORY BEHIND THE MUSIC AND LYRICS

MARK BRAKE with JEFF BRAKE

Skyhorse Publishing, Inc.

Copyright © 2025 by Mark Brake and Jeff Brake

All rights reserved. No part of this book may be reproduced in any manner without the express written consent of the publisher, except in the case of brief excerpts in critical reviews or articles. All inquiries should be addressed to Skyhorse Publishing, 307 West 36th Street, 11th Floor, New York, NY 10018.

Skyhorse Publishing books may be purchased in bulk at special discounts for sales promotion, corporate gifts, fund-raising, or educational purposes. Special editions can also be created to specifications. For details, contact the Special Sales Department, Skyhorse Publishing, 307 West 36th Street, 11th Floor, New York, NY 10018 or info@skyhorsepublishing.com.

Skyhorse® and Skyhorse Publishing® are registered trademarks of Skyhorse Publishing, Inc.®, a Delaware corporation.

Visit our website at www.skyhorsepublishing.com.

Please follow our publisher Tony Lyons on Instagram @tonylyonsisuncertain

10 9 8 7 6 5 4 3 2 1

Library of Congress Control Number: 2025933327

Cover design by Kai Texel
Cover artwork and images by Getty Images, Rosi Thornton, FreePNGimg.com, The Graphics Fairy, and Wikimedia Commons

Print ISBN: 978-1-5107-7821-4
Ebook ISBN: 978-1-5107-7822-1

Printed in the United States of America

Fox was the only living man. There was no Earth. The Water was everywhere.

"What shall I do?" Fox asked himself. He began to sing in order to find out.

"I would like to meet somebody," he sang to the Sky.

Then he met Coyote.

"I thought I was going to meet someone," Fox said.

"Where are you going?" Coyote asked.

"I've been wandering all over, trying to find someone. I was worried there for a while."

"Well, it's better for two people to go together . . . that's what they always say."

"Okay. But what will we do?"

"I don't know."

"I got it! Let's try to make a world."

"And how are we going to do that?" Coyote asked.

"Sing!" said Fox.

—Jaime de Angulo, *Indian Tales* (1953)

I always set out to write a children's book. I always wanted to write Alice in Wonderland. *I think I still have that as a secret ambition. And I think I will do it when I'm older.*

—John Lennon (1980)

*This book is dedicated to Cariad Jones, and
the memory of our late mother, Margaret, with love*

*The Beatles rendered in an early 1965 kindergarten book by Mark Brake.
Readers should note that, while the author hasn't yet developed an ability either
to spell with accuracy or to color between the lines (no Picasso, he), kudos is
nonetheless due for realizing that Paul McCartney is left-handed.*

Table of Contents

Introduction	xi
Please Please Me (1963): Live and Unleashed!	1
With The Beatles (1964): Beatlemania!	35
A Hard Day's Night (1964): The Fab Four Go Global	55
Beatles For Sale (1964): The Start of Studio Experimentation	73
Help! (1965): Riding So High	85
Rubber Soul (1965): The Album as Art Form	97
Revolver (1966): Psychedelia!	111
Sgt. Pepper's Lonely Hearts Club Band (1967): Art Rock	133
Magical Mystery Tour (1967): Experimental Rock	159
The Beatles (White Album) (1968): Postmodern Perfection	177
Yellow Submarine (1969): Full Steam Ahead	213
Abbey Road (1969): And in the End	219
Let It Be (1970): "Get Back" to Basics	235
Afterword	245

Introduction

"Study the science of art. Study the art of science. Develop your senses—especially learn how to see. Realize that everything connects to everything else."
—Leonardo Da Vinci, *An Exhibition of His Scientific Achievements* (Philadelphia, 1951)

Record-Making Revolutionaries

Why have we written this science book about The Beatles? Simply this: to pay tribute to history's greatest architects in the writing, recording, and artistic presentation of popular music.

The Beatles revolutionized many aspects of music and are often promoted as pioneers of 1960s youth and sociocultural developments. If sales and statistics are as important as aesthetics, then The Beatles triumph here, too. They remain the best-selling music act in history, with estimated global sales of 600 million units. The band are the most successful act on the US *Billboard* charts. They also hold the record for most number one albums on the UK albums chart, with fifteen. It's the same story in the singles charts—the band have the most number one hits on the US *Billboard* Hot 100 chart, with twenty, and the most singles sold in the UK, 21.9 million and counting.

Over the years, The Beatles have been recipients of many accolades. The list includes an Academy Award for Best Original Song Score for the 1970 documentary film *Let It Be*; seven Grammy Awards and four Brit Awards; and fifteen Ivor Novello Awards for their songwriting. The band was inducted into the Rock & Roll Hall of Fame in 1988, and subsequently each individual band member was inducted between 1994 and 2015. In June of 2010, Paul McCartney was presented with the Library of Congress Gershwin Prize for Popular Song. In his commemorative address at the ceremony, President Obama declared that

Introduction

"it's fitting that the Library has chosen to present this year's Gershwin Prize for Popular Song to a man whose father played Gershwin compositions for him on the piano; a man who grew up to become the most successful songwriter in history. The Beatles ... blew the walls down for everybody else. In a few short years, they had changed the way that we listened to music, thought about music and performed music forever. They helped to lay the soundtrack for an entire generation—an era of endless possibility and of great change." Moreover, Obama remarked that McCartney had "composed hundreds of songs over the years—with John Lennon, with others, or on his own. Nearly 200 of those songs made the charts—think about that—and stayed on the charts for a cumulative total of thirty-two years. And his gifts have touched billions of lives." In both 2004 and 2011, The Beatles topped *Rolling Stone*'s list of the greatest popular music artists in history. And finally, for now, *Time* magazine named the band among the twentieth century's hundred most important people.

In November 2023, British BBC radio DJ Lauren Laverne was asked to put the new Beatles single, *Now And Then*, in its historical perspective. Laverne replied, "Each fan has their own individual, very personal, relationship with The Beatles and their music. But then, collectively, they're part of our story as a country, as a culture. I mean it's not just about music; it's about modern Britain, about who we are. When we tell our story, we often use The Beatles to do it. It's one of the greatest stories of the past century. And it is timeless."

Laverne's point wasn't lost on British-Irish film director Danny Boyle. Responsible for movies like *Trainspotting, 28 Days Later,* and the Oscar-winning *Slumdog Millionaire,* Boyle was called upon by the London 2012 Olympics organizers to direct their opening ceremony. London didn't have the budget to plan a ceremony like 2008's opening in Beijing, so they had to rely on something else—a celebration of the history of British creativity, under the direction of Boyle himself. So, he set to work crafting a ceremony that would define both Britain as a country and the upcoming games. The result was one of the greatest events of performance art in history. The "pandemonium" section featured seven smoking chimney stacks representing the world's first Industrial Revolution, the women's suffrage movement, and two world wars, culminating in the 1960s and The Beatles as they appeared on the cover of *Sgt. Pepper's Lonely Hearts Club Band*—the making of modern Britain.

Danny Boyle would revisit the cultural impact of The Beatles in his 2019 movie, *Yesterday*, the romantic comedy written by Richard Curtis. Directed with the usual dash and delight by Boyle, this counterfactual film asks "what if The Beatles hadn't existed?" Struggling fictional musician Jack Malik suddenly finds himself the only person who remembers the band and becomes famous for performing their songs. Imagine no Beatles. No *Sgt. Pepper*, no *Walrus*, no *Strawberry Fields*, no *Yesterday*. Imagine no *Imagine*, no best-ever Bond theme in *Live and Let Die*, no Concert for Bangladesh to inspire *Live Aid*, no stadium rock, and no Monty Python's *Life of Brian*. And that's just for starters.

As the film unfolds, and through a simple sci-fi trick of altering the "space-time-reality-consciousness continuum," we can pretend to hear Beatles music for the first time, to hear the band afresh by proxy, by vicariously being placed in the position of musical novices, quite unaware of their revolutionary impact. It's an ingenious thought experiment that carries considerable emotional charge.

"The Beatles": Clouded in Mystery

The aim of this book is to explain, at least in part, not only why The Beatles were so successful, but also why their success has such longevity, continuing to capture the imagination of minds in the twenty-first century.

When Paul McCartney published his two-volume book *The Lyrics: 1956 to the Present* in 2021, he helpfully explained the origin of the band's name: "Buddy Holly came along when we were fifteen or sixteen. Buddy's look fitted with the fact that John [Lennon] wore glasses. John had a perfect reason to pull his glasses back out of his pocket and put them on. Buddy Holly was also a writer, a lead guitarist and a singer. Elvis wasn't a writer or a lead guitarist; he was just a singer.... So, Buddy had it all. And the name, The Crickets. We also wanted something with a dual meaning." McCartney says that the actual origin of the name "The Beatles" is clouded in mystery, but "if we could find an insect that also had some double meaning ... now that it's been around awhile, you totally accept it."

Meanwhile, and less mysteriously, in a phone interview in February of 1964, George Harrison explained, "We were thinking of a name a long time ago, and we were just wracking our brains for names, and John came up with this name Beatles, and it was good, because it was sort of the insect, and also the pun, you know, b-e-a-t on the beat. We just liked the name, and we kept it."

Lennon had another, slightly different, recollection. In an August 1971 interview with Peter McCabe, quoted in *John Lennon: For The Record* and later aired in the radio documentary *John Lennon: The Lost Tapes*, Lennon is quoted as reminiscing about the origin of the band's name: "They asked me to write the story of The Beatles.... So, I wrote this: 'There was a certain man, and he came ... ' I was still doing like from school, all this imitation Bible.... 'And man came on a flaming pie from the sky and said you are Beatles with an *a*.'"

The Beatles and Remix Culture

As we log how The Beatles' sonic innovation evolved in the pages ahead, it's worth bearing in mind the idea of the band as an example of "remix culture." This is a term that describes a culture or subculture which endorses derivative art by combining or editing existing materials to make new creative works. Indeed, the word "remix" originally referred to music emerging in the late twentieth century during the heyday of hip hop. But remixing didn't begin with hip hop.

Looking back, we can see that seminal British rock bands, such as The Beatles and The Rolling Stones, became hugely successful for "remixing" older American music, as Elvis before them became world famous for remixing source material that was drawn from traditional black blues musicians years before. In the early 1970s, British rock band Led Zeppelin became hugely famous for innovating a new kind of incredibly loud electric blues and, within just a few years, became the biggest band on the planet. But Led Zeppelin also "remixed." Much of their source material was drawn from traditional Black blues musicians years before. All these musicians were simply doing what artists do. Copying from others, transforming these ideas, and combining them with other ideas to create a new synthesis.

Remix is homage, not appropriation. Artists have been sampling and remixing for centuries. Consider Pablo Picasso's *Les Demoiselles d'Avignon*. Painted in 1907, the picture portrays five nude female prostitutes in a Barcelona brothel. Each prostitute is shown in an unsettling and confrontational manner, while none is conventionally feminine. In fact, the prostitutes are slightly menacing and rendered with angular and disjointed body shapes. *Les Demoiselles d'Avignon* is thought to be seminal in the early development of both cubism and modern art. It was considered the gold standard for creativity because it was thought to be unprecedented; nobody had seen anything like it before. But when you dig a little deeper into Picasso's famous painting, the signs of remix are clear. The figure on the left shows facial features and dress of Egyptian or southern Asian style. The two adjacent figures are depicted in the Iberian style of Picasso's native Spain, and the two on the right are portrayed with African mask-like features. Indeed, according to Picasso, the ethnic primitivism evoked in these masks, inspired him to liberate "an utterly original artistic style of compelling, even savage force." The Picasso example serves to show that beneath the myths of creativity lies a more profound reality of remix.

Now consider the remix culture of early Beatles music. On their first five albums of 1963 to 1965, almost all of the covers were originally recorded by Black American soul artists. The city port of Liverpool is pivotal here. It's no coincidence that a band from Liverpool was so heavily influenced by American music made by Black people, given that culturally open Liverpudlian youths could ask sailors who were family or friends to buy Black records from overseas—records they could never buy in Britain or could only hear on the wireless. Local bands would fight to be the first to learn whatever records they'd got from their sailor contacts, and once performed, they'd lay claim to the song. The Beatles were one of the few Liverpool bands to clearly state at their gigs which Black artists had originally recorded the song they were covering, even before they had a record contract of their own. This practice brought attention to Black artists, inspiring young people from Liverpool to ask sailor friends to buy records by the original artist when traveling overseas.

The Beatles' Irish Roots

There's another strong, and more local, cultural influence in The Beatles' music. Liverpool is different from the rest of England. From the welcoming, friendly nature of its people to their accents, boasted to be a golden amalgam of Irish, Welsh, and visiting Scandinavian sailors, Liverpool feels distinct. Many local "Scousers" regard Liverpool as a city-state, separate from the country in which it just happens to sit. And many locals don't identify with England at all. Liverpool football fans are known to display banners which declare SCOUSE NOT ENGLISH and to boo the British national anthem at soccer games. One may ask what this distinction is all about, and how the city which made The Beatles is so different to the rest of Britain.

The answer lies in Liverpool's roots. An estimated three-quarters of its inhabitants have Irish ancestry. Irish immigrants streamed into the city after the 1798 Irish Rebellion, a well-known insurrection against the British Crown. Another period between 1845 and 1852 became known as the Great Famine, when Ireland suffered a period of starvation, disease, and emigration, and the British government decided to shut down its centrally organized relief efforts in 1847, long before the famine ended. These two events greatly impacted Liverpool's demographic make-up. So, it's hardly surprising that all four Beatles have ancestry in Ireland. Indeed in 1963, Lennon stated in an interview upon landing at Dublin Airport, "We're all Irish!"

Paul McCartney has Irish ancestry in both his father's and his mother's line. For example, his maternal grandfather, Owen Mohan, was born in Tullynamalrow, County Monaghan. It's not yet been established from where in Ireland the McCartneys hail, but either his grandfather or great grandfather emigrated to Scotland from Ireland before later settling in Liverpool.

When McCartney published *The Lyrics*, the book's editor, Pulitzer Prize-winning Irish poet Paul Muldoon, said he and McCartney really clicked because of their Irish roots.

"We were raised in similar ways," Muldoon said in an interview with *Irish News*. "I don't think there was that much religion in his house but there was some Christian and more specifically Catholic iconography in some of the songs . . . The Beatles' classic *Let It Be* is the key example of the Catholic influence because it is about resignation, that is also a very Catholic worldview; here we all are in this veil of tears, get used to it. . . . Of course Mother Mary has a Catholic feel, she is honored in the Catholic tradition in a way she is not in the Protestant faith. . . . Quite frankly he is very conscious of his Irish roots, his family setting seems to be one that would be recognizable to many Irish people. It sounds like it was a party house or a *cèilidh* house with someone often playing the piano, having a drink or telling a story."

Contentiously, Muldoon also suggests that, in a very real way, The Beatles invented Liverpool. It shot up from being a grim industrial city to one of the most famous places on the planet thanks to the band. In Muldoon's words, "What comes to mind, is a line from Oscar Wilde: 'There were no fogs before Dickens, no sunsets before Turner.'

Introduction

To extend that idea, there wasn't a Dublin before Joyce. These artists invent the place. In some sense, Van Morrison invented Belfast writing about Paris buns and Fitzroy Avenue, and in a strange way The Beatles invented Liverpool . . . among other things."

George Harrison, too, had a solid link to Ireland through his grandparents, the Frenches, who hailed from County Wexford in Ireland. The French family line traces back to thirteenth century Norman knights with the name of Ffrench (the second 'f' was later dropped), who settled in County Wexford at the time of William the Conqueror. George had abiding strong connections to his Irish cousins, visiting them in Drumcondra near Dublin as late as 2001, just before his passing.

And while Ringo's links to Ireland are the least obvious (his deep English ancestry dates back three generations on both sides), Ringo's maternal great-grandmother, Elizabeth "Minnie" Cunningham, was born in County Down in 1851, the same county in which Lennon's ancestry originates.

At the time of the release of his solo single, *The Luck of the Irish*, written and released in the summer of 1972, John Lennon spoke about the Irish and the city of Liverpool and their influence on his art:

"Liverpool is where the Irish came when they ran out of potatoes, and it's where Black people were left or worked as slaves. It was a very poor city, and tough. The people have a sense of humor because they are in so much pain, so they are always cracking jokes. They are very witty. And we talk through our noses. I suppose it's adenoids. I'm a quarter-Irish and long, long before the trouble started, I told Yoko that's where we're going to retire, and I took her to Ireland. We went around Ireland a bit and we stayed in Ireland and we had a sort-of second honeymoon there. So I was completely involved in Ireland."

Meanwhile, the Lennon name is an anglicized derivative of the Irish O'Lennon, which is a descendant of the ancient Gaelic Ó Leannáin clans. In ancient Celtic legend, the stag in the Lennon crest implies spiritual guides or priests. The Gaelic meaning of Ó Leannáin is "love," entirely apt for a writer who delivered a message of love to the world in songs like *All You Need Is Love*, and a musician who was a member of a band whose most common lyric was the word "love," used 312 times in 42 percent of their songs. In searching for information about his family history in 1974, Lennon found the name O'Lennon along with a declaration that "no person of the name Lennon has distinguished himself in the political, military, or cultural life of Ireland (or England for that matter)." Lennon quoted this passage on the booklet included inside his 1974 *Walls and Bridges* album, along with the typical amusing comment of, "Oh yeh?"

Hamburg in the Remix

Another influence on The Beatles as a cultural and musical remix force was Hamburg. As Lennon once said, "I was born in Liverpool, but I grew up in Hamburg." For, in the tough conditions in the live music clubs of the working-class St. Pauli district of the city, a red-light district home to strip clubs and raucous seamen's pubs, The Beatles were told to *"mach Schau"* ("put on a show") to attract more clientele. It was here that they developed into a tight musical unit with a distinctive musical style, honed their performance skills, and earned themselves a growing reputation as a live band. Grueling sets of up to eight hours every day meant that the band developed a huge repertoire of songs alongside their own original songs. They became a "human jukebox," a fact still in evidence in Peter Jackson's documentary *The Beatles: Get Back*, originally recorded in 1969. The sheer number of covers the band knew intimately is truly staggering.

McCartney told American producer Rick Rubin at the very start of the television miniseries *McCartney 3, 2, 1* that The Beatles had played around ten thousand hours during their Hamburg stint, following which they were no longer the "bum group" that had left Liverpool, but were now a highly professional act. As artist and musician Billy Childish said about The Beatles in Hamburg, "This is The Beatles doing what they did best: stomping out great rock 'n' roll music to a half-despondent audience of Reeperbahn sex freaks. Lennon was of the opinion that The Beatles' early performances were never matched in the recording studio, and I reckon he was right. [Their music was] full of Brylcreem, winkle-pickers, punk rock vigor, and is guaranteed to lift the heart of anybody who is alive and open to the essence of good music: raw sound and fun."

"If it hadn't been for Hamburg, there would be no Beatles," remarked their first manager, Allan Williams, in an interview in 1995. Williams continued, "The work there was so fantastically hard. And people would say to me: 'Allan, tell us the secret of how to be a Beatle.' I say: go to Germany for six months, work seven nights a week, eight hours a day. And then come back and ask me the same question."

In short, The Beatles had a great working-class work ethic. They returned to Liverpool a completely different band, unrecognizable from the one that had left.

"Up to Hamburg we'd thought we were okay, but not good enough. It was only back in Liverpool that we realized the difference and saw what had happened to us while everyone else was playing Cliff Richard shit," Lennon remarked in *The Beatles Anthology*. That difference was clear to him: "We were performers in Liverpool, Hamburg, and around the dancehalls. What we generated was fantastic when we played straight rock. And there was no one to touch us in Britain."

Science as Tradecraft

So, what do we mean by the "science" of The Beatles? Sure, as McCartney wrote in his song "Maxwell's Silver Hammer," Joan may be quizzical, studying pata-physical science at home, and Maxwell Edison may well be majoring in medicine. But that's clearly not enough to fill a book. Unimaginative souls may be used to thinking about science in a rather one-dimensional way. Science is something only practiced by white middle-aged males in lab coats, they believe.

Just a matter of test tubes, tachometers, and telescopes. But, in truth, science is far more complex than that.

When we think just a little longer and deeper about science, we can quite easily see that science has a number of chief characteristics. For one thing, science is an *institution*, from the Academy of Plato in classical Athens, to London's Royal Society, the oldest continuously existing scientific academy in the world. (The Beatles, too, are an institution!) Science is also a cumulative *body of knowledge*, one which is verified by direct and repeatable experiments in the material world. Science is a key *driver of the economy* too, and should anyone doubt this fact, just take a look at how far humans have come since science and trade were bound together in the early days of the Enlightenment. Science is also a *worldview*, one of the most powerful influences shaping human beliefs and attitudes to our planet which drifts across the universe.

But, most importantly for this book, science is a *method*, a tradecraft. It's not a fixed thing; it's a growing process. It's made up of a number of operations; some mental, some manual. Science is a discipline whose concern is with how things are done. First, you have a look at the job, recording music in a particular way, for example, then you try something and see if it will work. In the learned language of science, we begin with observations and follow with experimentation. Artists, such as The Beatles, observe in order to transform, through their own experience and feeling, what they see into some new and evocative creation. The apparatus of their craft, the tools with which they carry out their experimentation, are the musical instruments and the evolving recording studio. The apparatus is not particularly mysterious. And yet, these expert musicians took these tools of ordinary life and turned them to very special purposes.

When we think of science in this practical way, we can identify its functions and how they relate to technique. Science is about action, as well as fact. Its origin and development lie in its concern with techniques to provide for human needs. Like music. The mode of science is indicative; it shows us how to do what we want to do. The artistic mode makes us want to do one thing rather than another. Make a sound like a thousand monks on a mountaintop, maybe, or conjure up the cacophony of living in a lemony submersible. Neither the scientific nor the artistic mode is complete without the other and, in fact, neither in science nor in art is one to be found without the other.

It helps greatly to think of science as a *recipe*. A recipe tells you how to carry out certain tasks, should you need to do them. From this practical perspective, science is not merely a matter of thought. It is thought constantly translated into practice, constantly refined by technique. History shows how new aspects of science continually developed in this way. In the Stone Age, for example, even something as basic as an axe had to be hewn using the technique of teasing ground stone into a tool that fulfilled the needs of primitive humans. Such techniques marked the start of science itself. And, ever since, science can be seen as an evolution of recipes which used practical techniques to carry out desired tasks.

So, this book is a case study of The Beatles' tradecraft and their master technician, George Martin. A story about the evolution of their recipes, through practice, experimentation, and testing, for the writing, recording, and artistic presentation of popular music, the dissemination of which became a crucial dynamic in the shaping of contemporary culture.

Masters of Flow

One of the many magical things about The Beatles is that they were able to create such revolutionary music in such a remarkably short space of time. Their entire standardized catalog was recorded in just *seven years*. This catalog isn't just remarkable for the *amount* of music they made, the frequency of hit singles and albums, but also for the sheer *invention* of their recordings. And to fulfill this ambition, the band needed a work ethic unlike any other contemporary artists. Their first album, *Please Please Me*, was recorded in a single day. The crucial thing about George Martin capturing the immediacy of the band in this way, like taking a snapshot photo, is that there's a certain sublime magic to it. The spirit of the band got onto the grooves of the vinyl. As Steve Jobs famously once said, "Real artists ship." Many artists have ideas, but only the greatest of them deliver. So *Please Please Me* instilled a vital ethos in the band; work rapidly and deliver.

By the midships of their career, the band was taking more time with their recordings, and using the studio itself as an instrument. They were creating albums which weren't simply a collection of songs, but works of art. Again, this ambition led to the discovery of new concepts and creative solutions that no one else had tried before. And yet, they never lost sight of the adage that real artists ship. Consider the song many regard as the greatest popular recording of all time: "A Day In The Life." Lennon had composed most of the song but was missing a middle eight (a varied section of a song that usually appears after a second chorus to break it up). McCartney suggested a solution which, though less than ideal (different tempo, sightly jazzy), Lennon said, "That'll do." Thus, one of the greatest rock and pop recordings is the product of a "get things done" attitude.

And so, The Beatles were masters of working in a state of flow. American drummer Vinnie Colaiuta is credited with the maxim, "Thought is the enemy of flow." Overthinking your compositional creative process and its recording takes the artist out of their state of flow. Hungarian-American psychologist Mihaly Csikszentmihalyi has more to say about flow, being "in the zone," the trance-like altered state of total absorption and effortless concentration. Csikszentmihalyi maintains that, when artists are fully engaged in their work, they experience a kind of ecstasy, with little need for thought, where the tracking of time is lost and the music "just flows out."

The Beatles were also rule breakers. Perhaps the most famous rule breakers in all of music history. The songwriting partnership of

Lennon and McCartney is the most successful of all time. And yet, amazingly, they did this without learning to read or write music. No formal tuition, no music theory. If an ingredient of their musical recipe wasn't meant to be there, but sounded good, they went with the flow as they didn't care for the "rules." And it meant that, without musical notation, their songs had to be good enough to be remembered by the band when they got into the recording studio. Such was the magic of their creative process. And to chart the development of all this, we shall tell the tale by reference to their studio albums.

Standardized Studio Albums

Given the structure of this book is based on The Beatles' studio albums, it is well worth noting here that those albums have been standardized globally since the first release of their music onto CD during 1987 and 1988. That standard is the following catalog of recording work mostly based around the album dates in the UK, the center of the Beatle universe:

- *Please Please Me* (1963; original UK album)
- *With The Beatles* (1963; original UK album)
- *A Hard Day's Night* (1964; original UK album)
- *Beatles For Sale* (1964; original UK album)
- *Help!* (1965; original UK album)
- *Rubber Soul* (1965; original UK album)
- *Revolver* (1966; original UK album)
- *Sgt. Pepper's Lonely Hearts Club Band* (1967; original UK and US album)
- *Magical Mystery Tour* (1967; original US album)
- *The Beatles ('The White Album')* (1968; original UK and US album)
- *Yellow Submarine* (1969; original UK and US album)
- *Abbey Road* (1969; original UK and US album)
- *Let It Be* (1970; original UK and US album)

So buckle up for a magical mystery tour of one of the greatest stories ever told, in the most extraordinary decade of the twentieth century. How four boys from Liverpool went from an underground club to conquering the planet. In just eight years, The Beatles caused a revolution, not just in music, but in society too, as figureheads for a blossoming youth counterculture, bringing new social, sexual, and artistic ideas into the mainstream.

They paved the way for every other artist that followed in their wake. Theirs was an influence that penetrated to the very heart of the post-war world. They were agents of change. Through their peaceful revolution, they became the undisputed voice of a generation. The most commercial band on earth, but also the most avant-garde and experimental. With groundbreaking sounds and science, trendsetting fashions, and an abiding sense of fun, this is the band that ripped up the rule book and transformed popular music and culture overnight. Without doubt, the best way to read this book is as a read-along companion to the songs themselves. And so, dear reader, let the Beatles music play.

The Science of
THE BEATLES

PLEASE PLEASE ME (1963)
Live and Unleashed!

"Love Me Do was written in one of our sessions at 20 Forthlin Road [Liverpool], up a little garden path, past my dad's lavender hedge, up by the front door where he had planted a mountain ash, which was his favorite tree ... John came up with this riff, the little harmonica riff. It's so simple. There's nothing to it; it's a will-o'-the-wisp song. But there's a terrific sense of longing in the bridge which, combined with that harmonica, touches the soul in some way."
—Paul McCartney, *The Lyrics: 1956 to the Present* (2021)

Please Please Me	Released: March 22, 1963	Recorded: September 11, 1962 - February 20, 1963	Duration: 31:59
Producer: George Martin	Studio: EMI, London	Label: Parlophone	Tracks: 14

Track Listing

Side One

No.	Title	Lead Vocals	Length
1	"I Saw Her Standing There"	McCartney	2:52
2	"Misery"	Lennon and McCartney	1:47
3	"Anna (Go To Him)"	Lennon	2:54
4	"Chains"	Harrison	2:23
5	"Boys"	Starr	2:24
6	"Ask Me Why"	Lennon	2:24
7	"Please Please Me"	Lennon and McCartney	2:00

Side Two

8	"Love Me Do"	McCartney and Lennon	2:19
9	"P.S. I Love You"	McCartney	2:02
10	"Baby It's You"	Lennon	2:35

(Continued)

Please Please Me (1963)

11	"Do You Want To Know A Secret"	Harrison	1:56
12	"A Taste Of Honey"	McCartney	2:01
13	"There's A Place"	Lennon and McCartney	1:49
14	"Twist And Shout"	Lennon	2:33

All songs written by McCartney-Lennon except: track 3, written by Arthur Alexander; track 4, written by Gerry Goffin and Carole King; track 5, written by Luther Dixon and Wes Farrell; track 10, written by Mack David, Barney Williams, and Burt Bacharach; track 12, written by Bobby Scott and Ric Marlow; and track 14, written by Phil Medley and Bert Russell.

Musical Archaeology

On April 3, 2023, the BBC reported an exciting new musical discovery. Almost sixty years after it was first made, the earliest known full recording of The Beatles playing a live concert in their homeland had suddenly surfaced. The recording captures the unleashed live sound of new music on the cusp of a breakthrough, of four musicians who were about to become the biggest band on the planet. In history. The sixty-minute tape recording was made by the then-fifteen-year-old John Bloomfield, a pupil at Stowe boarding school in Buckinghamshire, England, who had recorded the band when they played a concert at the school's theatre on April 4, 1963. Rather aptly, Bloomfield was a tech geek eager to try out his new reel-to-reel tape deck. And he told the BBC of the tape's existence when they traveled to Stowe to make a radio program about the sixtieth anniversary of the concert.

The gig is pulsating. Humans are rarely more thrilling than when they're under the pressure of performance. Electricity is crackling through their brains; cortisol and epinephrine are coursing through their veins. It's an awesome alchemy of sentience and biology—evolution at its freewheeling best. Performed to an almost entirely male audience, and despite hugely keen caterwauling and callouts, for once a live Beatles recording is not drowned out by the screams of the crowd. The concert truly captures the allure of the band's tightly polished but raw and energetic live show. The tape showcases their heady mash-up of a repertoire of R&B covers and Lennon and McCartney songs, kicking off with "I Saw Her Standing There" and segueing into Chuck Berry's "Too Much Monkey Business." The Beatles' debut album, *Please Please Me*, had been released just two weeks earlier, on March 22, 1963.

History is rich with discoveries which take us back in time, of course. And Bloomfield's discovery is no less significant for true connoisseurs of modern music. The Beatles, who tear zestfully through almost two dozen songs in the hour, made as much an impact on Bloomfield as they did on the rest of his generation, "I would say I grew up at that very instant. It sounds a bit of an exaggeration, but I realized this was something from a different planet."

The magic formula of The Beatles lay not just in the music. The band can be heard taking requests from the crowd, who call out the titles of songs that had been released to the public just a fortnight before. The tape's band/audience banter is archetypical. John can be heard doing joke voices. The huge popularity of Ringo is very apparent, and George had been singing so much of late that he'd lost his voice altogether. Despite the fact that Stowe was an all-boys school at the time of the recording, some girls had snuck in the back of the gig. As Bloomfield told the BBC, "It wasn't until they started playing that we heard the screaming, and we realized we were in the middle of Beatlemania. It was just something we'd never even vaguely experienced."

No Future in Show Business?

The Beatles' first single, *Love Me Do*, had been released in the UK on October 5, 1962, and stayed in that chart for the duration of the Cuban Missile Crisis, as the world worried the Cold War would escalate into full-scale nuclear war between the US and the USSR. The band's debut studio album, *Please Please Me*, was released roughly six months later in the UK on EMI's Parlophone label. (It is worth noting here that The

Please Please Me (1963)

Beatles' album releases in their home country were as the band, their manager Brian Epstein, and their record producer George Martin artistically *intended* them to be. In many other countries, their songs were seized by sloppy record labels who often mixed and matched tracks to profiteer from new collections, usually without the band's knowledge or consent. As stated in our Introduction, this book will use the official UK Beatles album discography, plus the US album *Magical Mystery Tour*, as this has since globally become the group's *de facto* catalog.)

The band had signed with EMI in May 1962, after famously being turned down by Decca Records, whose executives declared "guitar groups are on the way out," and that "The Beatles have no future in show business." Rarely has there been such spectacular stupidity in the history of recorded music. The band were instead signed to the Parlophone label run by George Martin. Impressed by The Beatles' raw energy, charisma, and musical ability, Martin first suggested they record a live album. He'd achieved great success in 1961 recording the hugely popular *Beyond the Fringe* comedy revue (featuring Dudley Moore and Peter Cook) with a tape recorder secreted directly under the stage of London's Fortune Theater. In Martin's own words on the documentary of The Beatles' 1995 *Anthology*, "I had been up to the Cavern and I'd seen what they could do—I knew their repertoire, knew what they were able to perform." But Martin found that the band's home venue in Liverpool, the subterranean Cavern Club, with its concrete walls resounding as a natural echo chamber, was quite ill-suited for a live recording. The die was cast. Martin changed his plan to re-create the electricity and élan of The Beatles' live shows in his recording studio—a studio album, but with a live twist. As Martin explained, "I said, 'Let's record every song you've got. Come down and we'll whistle through them in a day.'"

The Fifth Beatle

George Martin's influence on what was to come was profound. Before meeting the band, with whom his name would become synonymous,

Martin had produced novelty and comedy records in the 1950s and early '60s as head of Parlophone. Martin had worked with the likes of British comedians Peter Sellers and Spike Milligan, whose program *The Goon Show* was broadcast by the BBC between 1951 and 1960. *The Goon Show* parodied aspects of entertainment, business, art, politics, the State, education, the British class structure, and film and fiction. And it was also the main inspiration that led to Monty Python and had a huge influence on The Beatles' collective sense of humor. The band's playful, parodic, and absurdly surreal humor was present from the get-go with their irreverent and madcap answers to silly questions at early press events (as we will see later) and vividly evident in Lennon's award-winning books, 1964's *In His Own Write* and 1965's *A Spaniard in the Works*. So, the band were delighted when they discovered that Martin had produced some comedy recordings with the Goons. And they specifically asked that Dick Lester be the director of their first film, *A Hard Day's Night*, as they were huge fans of *The Running, Jumping and Standing Still Film*, which Lester directed and starred Peter Sellers, Spike Milligan, and Australian actor Leo McKern, who later co-starred alongside them in the movie *Help!*

George Martin was a supreme musician, too. While a select number of musicians also snuck their way into a Beatles mix here and there (including Eric Clapton, Donovan, Mick Jagger, Brian Jones, Keith Moon, Graham Nash, Billy Preston, and Keith Richards), Martin played on thirty-seven songs recorded by the band—around fifteen percent of their total output. That number does not account for the songs for which Martin arranged and conducted orchestral parts while not playing himself.

Through his symbiosis with The Beatles, George Martin almost single-handedly reformulated a record producer's role in music. In the rarefied company of technicians like Phil Spector and Quincy Jones, Martin was to become one of those elite producers almost as famous as the musicians they recorded. As the chapters of this book unfold, it's worth remembering that, in the dozen years before he worked with The Beatles, and as well as the string of popular comedy records, Martin produced jazz albums and symphonic, chamber, and choral recordings.

As George Martin told the *New York Times* in 2003, "When I joined EMI, the criterion by which recordings were judged was their faithfulness to the original. If you made a recording that was so good that you couldn't tell the difference between the recording and the actual performance, that was the acme. And I questioned that. I thought, okay, we're all taking photographs of an existing event. But we don't have to make a photograph; we can paint. And that prompted me to experiment."

Martin was a technician to The Beatles' art. He was a humble soul who'd been trained as a classical pianist and oboist, the instrument used to great effect on "Penny Lane," among others. Over the years, Martin was always modest about his involvement in the band's success. He'd wave away any credit due himself by replying to interviewers that his own contribution was secondary to the songwriting genius of The Beatles. However, the band, for their part, knew that Martin had a virtually infallible ear for arrangements. His technical advice and studio-based scoring and editing gave many of the band's best recordings their signature sound.

"I have experience of working with [my dad] and the band, actually he [George Martin] always used to say 'the boys' used to push him with technology," Martin's son, Giles Martin, said to the BBC in an interview in November 2023, marking the release of the new single, "Now And Then." "They didn't believe there were any walls [boundaries] to anything they did. And that's the thing. We should never listen to technology; we should listen to music and song. But if we can [get] people to fall in love with songs by helping with technology . . . that's the thing."

And so, in a technical sense, they should not necessarily surprise us: those resonant strings on "Eleanor Rigby," those screeching seagull tape-loops on "Tomorrow Never Knows," or those playfully weird sound effects on "Yellow Submarine." Martin had, in a very strong sense, been there before regarding sound effects. What was unique was the inventive musical magic that The Beatles brought to the studio. When Martin died in 2016, Quincy Jones described him as "my musical brother" who "knew the secrets of our craft that so few know today."

Inevitably, Martin's work with The Beatles overshadowed his other achievements. And even though the mainstay of that work lasted only from 1962 to 1970, producing thirteen albums and twenty-two singles for the band, a body of work that totals less than ten hours, it nonetheless revolutionized popular music. It is also sometimes overlooked that, once The Beatles went their separate ways, from 1974 Martin as good as doubled their initial output, supervising and producing archival material from the band's live gigs, such as those at the BBC and the Hollywood Bowl, as well as unreleased studio recordings that revealed a huge amount about how the band actually worked.

Beatles Laboratory

In today's world of multitrack mixing and mastering, the idea of recording a full album in a single day seems like a pipe dream. The modern process is simply too complex. But not in the UK in 1963. Back in the day, songs were recorded live to a two-track BTR (British Tape Recorder) machine. These "modern" tape decks were developed in 1930s Germany, but weren't used in Britain until several AEG Magnetophon recorders were discovered after the end of WWII. In 1946, it was an Abbey Road engineer, Berth Jones, who journeyed to Berlin and helped adapt the technology for what became the BTR. And that two-track BTR meant scant opportunity for overdubs or refined edits.

Nonetheless, from the beginning, technology lay at the heart of The Beatles' musical revolution. Throughout their career, the band used many technological innovations to drive the creative use of

Please Please Me (1963)

sound in their musical development. From *Please Please Me* to *Let It Be* in 1970, they were always pushing the boundaries of what was technologically possible. And the constraints of recording technology placed upon them during *Please Please Me* was the BTR-2 machine that was first introduced into Abbey Road in 1954, almost a full decade before.

So *Please Please Me* was recorded live in Studio 2 at Abbey Road on this two-track BTR-2. Only two sessions were originally booked to record the album. The rigid schedules at Abbey Road dictated that the sessions took place between 10:00 a.m. and 1:00 p.m., and 2:30 p.m. and 5:30 p.m. However, a third session, between 7:30 p.m. and 10:45 p.m., was also added. What happened in the next 585 raw and energetic minutes has become the stuff of legend. Since the songs were recorded live, this presented few opportunities for correcting errors. And so, the finished result was the authentic live sound of The Beatles, with the amps and drums mic'd up with little or no sound isolation between the instruments. This raw sound was achieved by overdriving the Top Boost of their Vox AC30 amplifiers. As AC30 amps have a "clean" channel, the guitar sounds clean. But Top Boost gives a grittier sound, and overdrive cranks it up to the max. As George Harrison recalled in the multimedia retrospective project, *The Beatles Anthology*, "We were permanently on the edge. We ran through all the songs before we recorded anything."

By 10:30 a.m., the band were ready and primed with their, now iconic, equipment: McCartney's violin-shaped Hofner bass, Ringo's Premier drum kit, Harrison with his Gretsch Duo-Jet and a J-160E Gibson "Jumbo" acoustic, and Lennon with the same Gibson Jumbo and a Rickenbacker 325. With the technicians fully aware of the fact that, in 1963, Abbey Road sessions ran precisely to time, thus began a race against the clock.

Years of playing at clubs in Hamburg and relentless touring around Britain had prepared the band well for this grueling session. But their knowledge of the recording process was very limited.

"The Beatles didn't really have much say in recording operations," George Martin said later. "It was only after the first year that they started getting really interested in studio techniques."

The band always wanted to get the recording perfect, time allowing, so the songs were not all recorded in one take. They would listen to each take, and then do two or more takes until they were satisfied.

"This [first] album was one of the main ambitions in our lives," McCartney said in Keith Badman's 2009 book, *The Beatles: Off the Record*. "We felt that it would be a showcase for the group, and it was tremendously important for us that it sounded bang on the button. As it happened, we were pleased. If not, sore throats or not, we'd have done it all over again. That was the mood we were in. It was break or bust for us."

With only a twin track at their disposal, once the live performance of each song was captured on track one, the second track was used to capture vocal, percussion (hand claps), or instrument overdubs. It's here that George Martin's studio expertise came to the fore. In "A Taste Of Honey," Martin uses the technique of "double-tracking" where vocals are recorded twice, resulting in a far richer, rounder sound. As it's very tricky for a singer to replicate the same part in exactly the same way twice, the vocal part is recorded twice resulting in two different performances of the same part. This creates a fuller, "chorused" effect with double tracking. But, if one simply blends a single performance in perfect sync, the whole double tracking effect is lost. The Beatles would use this recording technique again and again on forthcoming albums.

On other songs, we see the first appearance on a Beatles record of the "vari-speed" recording technique, one of Martin's favorite tricks. Working on Parlophone's successful comedy records, Martin had picked up quite a few tricks of the trade. Having to work with two tracks was a constant source of frustration for him, especially when classical recordings at EMI were made with *four*-track technology. For non-classical recordings, additional instruments and vocal overdubs could only be achieved by "bouncing" the tracks to create valuable additional space. (The reason for this being the relative complexity of classical composition. The Beatles' innovative style meant they required more tracks than traditional pop bands had up until now.) The downside was that, with each bounce, there is potential for loss of sound quality. To counteract this, Martin tried to engineer something magical.

As EMI sound engineer Geoff Emerick noted, "That session [*Please Please Me*] was my first exposure to George Martin's signature 'wound-up' piano—piano recorded at half speed, in unison with guitar, but played an octave lower. The combination produced a kind of magical sound, and it was an insight into a new way of recording—the creation of new tones by combining instruments, and by playing them with the tape sped up or slowed down. George Martin had developed that sound years before I met him, and he used it on a lot of his records."

How was vari-speed recording used, for example, on the *Please Please Me* track "Misery"? Martin recorded piano at half-speed to enhance George Harrison's opening guitar chord, thereby adding a very appealing sound layer on top. As Emerick noted, "Overdubbing a half-speed piano is not the easiest thing to do, either, because when you're monitoring at half speed, it's hard to keep the rhythm steady. There certainly were more than a few expletives coming from George as he struggled to get the timing down while overdubbing onto the song 'Misery,' on both the spread chord that opens the song and on the little arpeggios and chord stabs that are played throughout." On "Misery," Martin also joined together the start of take seven and the end of take nine to create a final version.

The vari-speed technique would be used extensively by the band during their recording career. But, on *Please Please Me*, Martin himself was pleased enough with the results that he turned his attention to "Baby It's You," where he added a percussive bell instrument called a celeste to double-up on Harrison's guitar part. Emerick later recalled,

"Again, he was trying to get a new tone by blending the two instruments together—and, again, nobody had ever heard a sound like that before. Later on, he also tried adding some normal-speed piano to the song, but decided it wasn't necessary, so only the celeste made it to the record."

"At the end of the [album] recording, George Martin looked down from the control room and said in amazement, 'I don't know how you do it. We've been here recording all day and the longer you go on, the better you get!'" McCartney is quoted as recalling in *The Beatles: Off the Record*.

"Waiting to hear that LP played back was one of our most worrying experiences," Lennon is quoted as saying in *The Beatles Anthology*. "We're perfectionists: if it had come out any old way, we'd have wanted to do it all over again. As it happens, we were very happy with the result."

When they began to record the *Please Please Me* album, Martin and The Beatles had already had four tracks in the can. They were the two singles "Love Me Do" and "Please Please Me," and their respective B-sides, "Ask Me Why" and "P.S. I Love You." And this meant that the band had ten more songs to find, so they could complete the customary fourteen tracks of a British album.

At the time, it was tradition for artists to rely on professional songwriters, with even the likes of Buddy Holly, Chuck Berry, and Roy Orbison leaning on fellow songwriters to help fill their albums, but it's very much worth noting that, on their very first album, Lennon and McCartney soon had writing credits for a full *eight* of the fourteen tracks of *Please Please Me*—an unrivaled triumph for a popular band at the time. The insurrectionary arrival of a band who wrote most of their own material was a revolution in its own right. And history is testament to the huge number of later songwriters who cited and continue to cite The Beatles as the exemplary band whose do-it-yourself approach inspired them to set up bands of their own.

Please Please Me (1963)

Emphasizing the live nature of the *Please Please Me* studio session, Martin later commented, "It was a straightforward performance of their stage repertoire—a broadcast, more or less," a performance that was not unlike their regular live sessions on BBC radio. The Beatles' manager, Brian Epstein, had secured the band a day off from their exhausting touring commitments for the eve of the recording session so that the four lads would arrive prompt and fresh at EMI on the morning of February 11, 1963. But, as Scottish poet Robbie Burns once wrote, "The best laid schemes o' mice an' men, Gang aft a-gley." Rather than prompt and fresh, the band turned up late, and Lennon was nursing a bad cold. As studio engineer Norman Smith was later to recall, "[John's] voice was pretty shot." Pharmacology came to the rescue in the form of Zubes, an old-school English-made throat lozenge. While this might seem like an irrelevant medical detail, it goes some way to explaining Lennon's incredible vocals as we run through the album's track listing.

Track by Track: *Please Please Me*

"I Saw Her Standing There"	Recorded: February 11, 1963	McCartney *(vocals, bass guitar, handclaps)*	Lennon *(backing vocals, rhythm guitar, handclaps)*
(McCartney-Lennon)	UK Release: March 22, 1963	Harrison *(lead guitar, handclaps)*	Starr *(drums, handclaps)*

"I Saw Her Standing There" began as a work in progress in 1960 and found its final flourish at EMI Studios. In his 2021 book, *The Lyrics: 1956 to the Present*, McCartney said of the song, "I've written a lot of songs, but certain ones stand out, and if I had to choose what I thought was my best work over the years, I would probably include "I Saw Her Standing There." No, I would *definitely* include this one . . . I first played this song to John when he and I got together to smoke tea in my dad's pipe (and when I say tea, I *mean* tea.) . . . I don't know where 'beyond compare' came from, but it might have come out of [Shakespeare's] Sonnet 18: 'Shall I compare thee to a summer's day?' I may even have been conscious, as a child, of the Irish song tradition—of a woman being described as 'beyond compare.'"

(For example, the old ballad *Yon Green Valley* includes the lyrics, "I'll think no more of her yellow hair/Her two black eyes are *beyond compare*/Her cherry cheeks, and her flattering tongue/Twas what beguiled me when I was young.")

Now a rock-and-roll standard, "I Saw Her Standing There" is said to be, by those who knew the band in their early days, one of a couple of the most characteristic Liverpool club scene songs, the other being The Big Three's cover of Richie Barrett's "Some Other Guy," which The Beatles also played live. In a clear move to capture some of the excitement of The Beatles' live performance of the song, the track kicks off with McCartney's count off, "one-two-three-FAW!" which, prompted by Martin, is designed to give that vital feel so evocative of the band's gigs at The Cavern and The Casbah.

The track is electrifying, and beautifully sets up the live feel for the rest of the album. The song also lays to rest any doubt that listeners may have that the band were the "charismatic powerhouse" which rocked Merseyside in the early sixties. British musician Lemmy, of Hawkwind and Motörhead fame, recalled those early Cavern days in *Mojo* magazine in 2006:

> I was sixteen, living in Anglesey, the big seaside resort for Liverpool, when I saw The Beatles play at The Cavern. The Cavern wasn't licensed, so anybody could go. They used to have lunchtime shows as well, with three bands, and the secretaries would come in out of the offices with their hair in curlers under their scarves and dance around their handbags in their lunch hour. The Beatles was the most magical thing I've ever seen—this complete four-headed monster, each of them as good as the other. You

could see them feeding on each other's energy, and you could see they were having fun. They'd swear at the crowd and when they'd make a mistake they'd crack up laughing—which bands didn't do at all back then. I could never believe the fucking harmonies, they were so good: three possible vocals—four with Ringo, but Ringo wasn't with them then. This was 1961 or '62, before their first single had come out.

Already masters of the live performance, the band delivered a perfect version of "I Saw Her Standing There" on the initial take, singing and playing live, and capturing on record their soon-to-be-famous falsetto ooos!—that early mop-topped motif of Beatlemania. And yet, just to be sure, Martin called for a second take, which proved less successful due to a mix-up with the lyrics in the chorus.

No doubt convinced they'd cracked it on take one, McCartney ended the second take with a dispirited bass slide and an analytical Lennon declared, "Dreadful." After numerous other takes, and with patience growing thin, McCartney began take nine with added zeal, starting with a spat-out and strident count-in, which turned out so rousing that Martin later edited it onto the front of take one, thereby creating one of popular music's best intros since Elvis sang "Well, it's one for the money, two for the show" on his debut seven years before.

English composer Adem Ilhan in *Mojo* magazine of July 2006 reminisced on the power of the song:

> When I started playing the guitar, when I was ten or eleven, this was one of the first ones I played. The simplicity of it, chord-wise, hit me really vividly. I felt like a punk! It rocks, it changes exactly when you want it to change, every syllable just seems to fall in the right place. The rawness and foot-tapping swing-around-ness is not rivaled anywhere else. The way the count's left in, the guitar tone, the whole

atmosphere—you're in the room with them. Making records in a few days like this and having done with it, it's really a lost art.

Rather poignantly, "I Saw Her Standing There" was also the last song John Lennon played at a major live performance. The occasion was a concert with the Elton John Band at Madison Square Garden, New York, on November 28, 1974. In a manner typical to Lennon, he introduced the live performance of the song in the following irreverent manner: "I'd like to thank Elton and the boys for having me on tonight. We tried to think of a number to finish off with so I can get out of here and be sick, and we thought we'd do a number of an old, estranged fiancé of mine, called Paul. This is one I never sang, it's an old Beatle number, and we just about know it."

This live version of the song was subsequently released twice; once as the flip side to Elton's 1975 hit single, "Philadelphia Freedom," and secondly in its own right as a UK single in March 1981 after Lennon's untimely passing. It was the first time that a version of "I Saw Her Standing There" had entered the UK charts.

In conclusion, built around blues shifts and the band's beautifully discordant harmonies, "I Saw Her Standing There" dropped onto the British pop scene like an incendiary device. The shockwaves of the song's gutsy rawness affronted the then-dominant bland chart harmonies that had been influenced by Broadway. The revolution had begun.

"Misery"	Recorded: February 11, 1963	McCartney *(vocals, bass guitar)*	Lennon *(vocals, rhythm guitar)*
(McCartney-Lennon)	UK Release: March 22, 1963	Harrison *(lead guitar)*	Starr *(drums)*

In 1999, forthright British journalist Jeremy Paxman more than met his match when he interviewed British music legend, David Bowie. Regarded as one of the most influential musicians of the twentieth century, Bowie held forth on a variety of topics, including the role of drugs in creativity, the huge potential of the internet, and the true nature and identity of British rock music. Bowie's hypothesis about the self-conscious aesthetic of British rock was fascinating:

> We've always been good at music. We're not truly a rock nation. Everything we do in rock 'n' roll has a sense of irony attached to it. We know that we're not the Americans. We know it didn't spring from *our* souls. So, as the British always do, they try and do *something* with it, to make them feel smug. And that's what we're good at doing.

The imperious yet pastiche dance-posing of Mick Jagger. The outrageously facetious comedy-camp of Elton John. But with The Beatles, the goofing is on an entirely different scale of magnitude. The bangs, clangs, and "Captain! Captain!" of *Yellow Submarine*. The handclaps and "ho-ho-hos" of "Ob-La-Di, Ob-La-Da." The band leaving McCartney's "fucking hell" in the final mix of "Hey Jude" after he'd made a mistake on the piano and the lads thought it too funny to edit out. Or "Back In The USSR," which jokingly and lovingly subverts the narrow, patriotic sentiments of the Beach Boys' "California Girls" and Chuck Berry's "Back in the USA." When director Peter Jackson's documentary series, *The Beatles: Get Back,* dropped on Disney+ in November of 2021, many observers remarked on the extent to which the band "joked around," even during the apparently stressful sessions that culminated in the *Let It Be* album, with folk often amazed at the level of "silliness." American musician and record producer Rick Beato commented that "what was so amazing about this [the *Get Back* movie] you see how much they goofed around. . . . They knew so many cover songs, from being a band since the mid to late fifties, they could just break into songs, they would do hilarious things. . . . They would just do these gags, over and over. I would say three quarters of the film is them goofing around."

As McCartney himself described this to Rick Rubin in *McCartney 3, 2, 1*: "A really good thing that did happen: we all *knew* we had the freedom to goof around." So, even during the band's early mop-top days, it was quite common for the band to dissolve into unrestrained laughing fits.

Witness "Misery." According to British writer Ian MacDonald, "Misery" is a "droll portrait of adolescent self-pity." And one must ask how much self-pity was really being felt by a bunch of young twenty-something males who had just experienced the debauchery of Hamburg? McCartney has gone on record to say that the band had only previously had sex with "normal" girls from Liverpool. When they traveled to Hamburg, however, a Pandora's Box of sexual possibility was pried open before them. For the only women who haunted the Hamburg clubs late at night were dancers, strippers, and prostitutes. George Harrison, then a mere seventeen years of age, dubbed Hamburg "the naughtiest city in the world" on *The Beatles Anthology* DVD. Moreover, also on the *Anthology*, McCartney commented, "By the time you got to Hamburg, a girlfriend there was likely to be a stripper, so to be suddenly involved with a hard-core striptease artist, who obviously knew a thing or two about sex . . . it was quite an eye-opener."

Meanwhile, in "Misery," and for the more respectable mainstream popular music market, the lads sing about a single lost love, "Can't she see she'll always be the only one?"! The picture becomes clearer when one knows that Lennon and McCartney initially wrote "Misery" for Helen Shapiro, with whom they had recently toured and who had recorded two 1961 UK chart toppers, "You Don't Know" and "Walkin' Back to Happiness," when she was a mere fourteen years old. And so, "Misery" was mostly a goof, with Lennon and McCartney singing together in a raffish manner through the parodic intent echoed in the jaunty, *Abba-esque*, but ultimately comic piano parts that George Martin added later, to the arch and fading *la-la-la-las* of Lennon. In short, and as David Bowie would have us believe, "Misery" is a consummate exercise in British self-mockery.

Please Please Me (1963)

"Anna (Go To Him)"	Recorded: February 11, 1963	McCartney *(backing vocals, bass guitar)*	Lennon *(vocals, acoustic rhythm guitar)*
(Alexander)	UK Release: March 22, 1963	Harrison *(backing vocals, lead guitar)*	Starr *(drums)*

In his 2013 book, *The Beatles: All These Years*, Mark Lewisohn quotes McCartney as saying, in 1987, that "If The Beatles wanted a sound, it was R&B. That's what we used to listen to and what we wanted to be like. Black, that was basically it. Arthur Alexander."

"Anna (Go to Him)," or just "Anna," is a tribute to that very Arthur Alexander, an Alabama country-soul pioneer, and one of Lennon's heroes. "Anna," an Alexander song, was originally recorded and released as a single by Dot Records in late 1962. As with "Misery," all is not what at first it seems with these early songs. Alexander's lyrics for "Anna" claim to be about the early days of his relationship with his partner Ann, and how her wealthy ex tried to win her back. But Richard Younger, Alexander's biographer, points out that in fact "it was surely Arthur who had been unfaithful in his marriage vows, [though] in the song he puts himself in the role of the abandoned lover." Artistic license, at best.

Contemporary male hypocrisy notwithstanding, a considerable portion of the *Please Please Me* studio session is The Beatles rattling through beloved covers of their Hamburg and Liverpool live sets. Songs that truly inspired the band. And, as is clear from McCartney's comments above, it's noteworthy that all the non-McCartney-Lennon songs on their first album had previously been recorded, and/or popularized, by black soul artists, confirming McCartney's claim of the band as an R&B outfit.

"Anna (Go to Him)" is the first Lennon-led cover of the session, and, given the band's rendition had been well-honed through constant professional performances, the recording was a breeze and "merely" a matter of replicating as good a live take as Martin could get onto tape.

According to the *Anthology*, George Harrison was also a huge fan of Alexander.

"I remember having several records by him, and John sang three or four of his songs. Arthur Alexander used a peculiar drum pattern, which we tried to copy, but we couldn't quite do it, so in the end, we invented something quite bizarre but equally original."

Harrison himself was responsible for playing on guitar the original song's piano intro, which was played by Floyd Cramer, an American pianist who became famous for his melodic signature playing style, which was a cornerstone of the pop-oriented Nashville sound of the 1950s and 1960s. All in all, "Anna (Go to Him)" was, given the band's polish from sheer practice and performance, done by take three.

"Chains"	Recorded: February 11, 1963	McCartney *(harmony vocals, bass guitar)*	Lennon *(harmony vocals, rhythm guitar, harmonica)*
(Goffin-King)	UK Release: March 22, 1963	Harrison *(vocals, lead guitar)*	Starr *(drums)*

"Chains" was penned by the famous husband-and-wife duo Gerry Goffin and Carole King, who were such a huge inspiration to the band's own songwriting partnership that Lennon once declared a wish that he and McCartney would become the "Goffin and King of England." The song was one of those that highlighted the band's ability to dig out American pop classics.

"Chains" was originally offered to The Everly Brothers who recorded their version in the summer of 1962, but chose not to release it. McCartney explained in the liner notes to *On Air – Live at the BBC Volume 2*: "With our manager, Brian Epstein, having a record shop, NEMS, we did have the opportunity to look around a bit more than the casual buyer."

Thus it was that The Beatles engaged in a kind of insider trading with their music sources. George Harrison was especially fond of the song, having bought a version of the record in late 1962 and bagging the lead vocals for himself. At the *Please Please Me* sessions, the band laid down a couple of complete takes of the song, with the first being chosen for the album.

"Boys"	Recorded: February 11, 1963	McCartney *(backing vocals, bass guitar)*	Lennon *(backing vocals, rhythm guitar)*
(Dixon-Farrell)	UK Release: March 11, 1963	Harrison *(backing vocals, lead guitar)*	Starr *(vocals, drums)*

Continuing the album's "live" feel, and with a very live sound in the studio, "Boys" features Ringo on lead vocals, as he did onstage, with the rest of the band backing him up with *bop, shoo-wop; bop-bop, shoo-wop*. Not only that, but Ringo played drums and sang at the same time (never an easy thing to do), and thoughts of Hamburg must have gotten a grip, as he aced it on the first attempt—the only track on the album to be wrapped in a single take. As Ringo later recalled, "We didn't rehearse for our first album. In my head, it was done 'live.' We did the songs through first, so they could get some sort of sound on each one. Then we had to just run, run, run them down."

We spoke earlier, during our comments on "Misery," about The Beatles' seemingly unending appetite for goofing around. "Boys" is another case in point. Again, in that typical British way of parodying "proper" rock with the sense of irony of which Bowie spoke, the band cover this Shirelles B-side without even changing the lyrics, with The Beatles playing at being a girl group. According to McCartney, "Boys" was "a fan favorite with the crowd. And it was great—though, if you think about it, here's us doing a song and it was really a girls' song. 'I talk about boys now!' Or it was a gay song. But we never even listened. It's just a great song."

Ringo had been singing "Boys" since his pre-Fab days in Rory Storm and the Hurricanes, which brings us nicely on to the topic of The Beatles as a democratic unit. The prevailing consensus at the time was that every pop group had to have a frontman: Buddy Holly and the Crickets, Bill Haley and the Comets, Frankie Valli and the Four Seasons; that kind of corny old thing. But George Martin, to the band's eternal benefit, rejected the idea of naming them First Name Surname and The Beatles. Thus it was that the notion of The Beatles as a unified collective was born. The band took Martin's idea further forward by giving each band member their own lead vocal spot on the album.

"Ask Me Why"	Recorded: June 6/ November 26, 1962	McCartney *(backing vocals, bass guitar)*	Lennon *(vocals, rhythm guitar)*
(McCartney-Lennon)	UK Release: January 11, 1963	Harrison *(lead guitar)*	Starr *(drums)*

Written in early 1962, "Ask Me Why" was mostly a John Lennon song. In his book *Paul McCartney: Many Years From Now,* Barry Miles quotes McCartney as saying, "It was John's original idea and we both sat down and wrote it together, just did a job on it. It was mostly John's." Already part of their live act in early 1962, "Ask Me Why" was one of the songs the band performed at their first Parlophone session at Abbey Road on June 6, 1962, with the band also performing the song for the BBC Light Programme's *Teenager's Turn – Here We Go* on June 15, 1962, recorded at the BBC Playhouse in Manchester.

Cast in the style of Smokey Robinson and the Miracles, an influence on Lennon, and with an opening guitar phrase remixed from the Miracles' "What's So Good About Goodbye," as American musicologist Alan Pollack points out, "Ask Me Why" is structurally complex—containing, for example, three different variants of the verse. According to Pollack, "This is just about the fussiest, most complicated form we've yet seen ... we have a strong presence here of chord streams, though this time the chords are jazzy." "Ask Me Why" has a live ending.

Please Please Me (1963)

"Please Please Me"	Recorded: September 11/ November 26, 1962	McCartney (*vocals, bass guitar*)	Lennon (*vocals, rhythm guitar, harmonica*)
(McCartney-Lennon)	UK Release: January 11, 1963	Harrison (*harmony vocals, lead guitar*)	Starr (*drums*)

Weird to report but "Please Please Me" appears to have been born out of Roy Orbison's "Only The Lonely." Penned at the house of Lennon's Aunt Mimi, the song had at first hoped to ape the vibrato melancholy of Orbison's impassioned pop song. But when the band first played "Please Please Me" for Martin, he told them the song sounded dreary and suggested that they pick up the tempo and add a simple harmonica introduction.

On wrapping up the song's recording, a chuffed Martin had engaged the studio's control room intercom: "Congratulations, gentlemen, you've just made your first number one." Kudos is no doubt due to Martin, given the man's background and age, in showing a unique acumen in recognizing the "shock of the new" in the band's musical brand. He was, naturally, modest about the matter.

"I was purely an interpreter. The genius was theirs, no doubt about that."

And yet, Martin's modesty disguises the fact that there was no other contemporary working studio technician accomplished and skilled enough to empathically produce The Beatles without taking too much away from what they already were. As a producer and technician, Martin was to curate and cultivate, in the most graciously English way, the band's genius at play.

Martin's suggestions transformed the song, and a fortnight after its release, "Please Please Me" was top of the UK singles chart's "hit parade"—The Beatles' first big hit—thereby fulfilling Martin's foresight.

Compared to the still extant tapes of The Quarrymen (the skiffle group formed by Lennon that evolved into The Beatles in 1960), the live Star Club recordings from Hamburg, and the couple of preceding

EMI sessions, "Please Please Me" provides an energized performance and an arrangement more complex than anything the band had tried before. More evidence that the creative and reflexive influence of Martin made its mark even at this early date.

The Everly Brothers also had clear influence on the track. From their live repertoire, Lennon and McCartney were accomplished Everly Brothers mimics.

"Last night I said these words to my girl/'I know you never even try, girl.'" Here, the verse barely strays from a single chord as Lennon sings, building to a climax in the chorus until the call-and-response of "Come on (come on)," between a raw Lennon on the one hand and a heartily harmonized McCartney and Harrison on the other, and resolves into a pleasing three part, "Please, please me, whoa-yeah, like I please you."

"I saw The Beatles live at Southend Odeon in 1963," British musician Wilko Johnson said in the Beatles special edition of *Mojo* magazine of July 2006. "I say 'saw' because it was almost impossible to hear the band. Their AC30 amps and the primitive house PA were no match for the tumult of screaming, shouting, whistling, and cheering—to which I contributed my fair share. It was madly exciting. "Please Please Me" was their first Number 1, and it just sounds like a hit record. It opens with an urgent, chiming figure on guitar and harmonica—an unusual instrument on a pop record at the time, particularly when played in such a funky way. Then the vocals come in—what a sound! Two great voices going full tilt in harmonies at once raw and beguilingly melodic, the like of which had never been heard before. The record is full of hooks—the guitar/harmonica passages; the 'come on, come on' choruses leading into the punning title line; the falsetto leaps (images of shaken mop-tops!) and everybody knew it was the best sound in the world, and it was British. I traded the Campagnolo gears from my bike for my ticket to that Odeon show. A schoolmate had queued all night to buy it, so the price was high—but I think I got the best of the bargain. Whenever I hear 'Please Please Me' I remember why."

Please Please Me (1963)

While not emotionally compelling in the same way as a Roy Orbison song, "Please Please Me" nonetheless grips the listener with its tale of a sexually frustrated Lennon who is imploring the girl to please him, just like he does her. Given what we know about the band's time in Hamburg, it's reasonable to assume Lennon is here referring to oral sex, but, naturally, such a suggestion would be categorically denied due to the band's very clean-cut image that Brian Epstein had conjured up at the time. Lennon was later to mischievously say of the song, "I was always intrigued by the double use of the word 'please.'"

"Love Me Do"	Recorded: June 6; September 4, 11, 1962	McCartney *(vocals, bass guitar)*	Lennon *(vocals, rhythm guitar, harmonica)*
(McCartney-Lennon)	UK Release: October 5, 1962	Harrison *(backing vocals, acoustic rhythm guitar)*	Starr *(drums)*

As McCartney wrote in *The Lyrics*: "The biggest influence on John and me was the Everly Brothers. To this day, I just think they're the greatest. And they were different. You'd heard barbershop quartets, you'd heard The Beverley Sisters—three girls—you'd heard all that. But just two guys, two good-looking guys? So, we idolized them. We wanted to be like them."

And so we have "Love Me Do," with its Everly Brothers-inspired vocal duet of Lennon and McCartney, the pared-down lyric of a blues composition, and the ungainly skiffle-esque beat. And yet, even here, the listener can sense the youthfully charged energy lurking just below the surface of the remix. From the get-go, the trademark sound is forged: Lennon's lead harmonica, the backing track of rhythm guitar, bass, and drums, with Paul and John singing their duet in Everly harmony. "Love Me Do" is pretty much an early and definitively quirky and cunning natural recording of the band, which introduces the world to the new sensation that is The Beatles.

Orchestrated under Martin's direction, the recording presents the band's two main protagonists, Lennon through his soulful harmonica riff and, when the music stops, McCartney through his unaccompanied phrasing of the song's title. Couple this with that capstone of a well-crafted song, the hook, which in this case is the bittersweet harmony of "Ple-e-e-ease" and the nod to Ringo with the hippodrome cymbal crash at the end of Lennon's solo, and you have the perfect Beatles recipe. Harrison is the silent one, strumming shyly in the background. But the subtlest effect was the record's air of unvarnished honesty. Martin chose a vocal harmony that oozed reverb but decided that the rest of the recording would be refreshingly "dry" when compared with the echo-obsessed sound of pop records in the few years before "Love Me Do." The net result was a directness that perfectly chimed with the band's candid image, distinguishing them from everyone else in sight.

So, once more the "live" feel is palpable on the recording; it sings to the listener's heart and ears. We feel the timorous tremble in McCartney's vocal, the faltering sound of his bass line, and likewise for the hesitancy of Ringo's drum part (though, it's worth noting that the looser, swinging style of the likes of Ringo and Charlie Watts of The Rolling Stones was only just coming into vogue.) We can aurally divine on this first official release of the band's just how much they want to please, how much of the lads' hopes rest on the success of the song, how the recording resonates with the fact the band is placing their "all" on the line, and the way in which they must have felt that they could make the future happen for themselves.

The outcome of the band's remix composition with Martin's studio expertise was a recording that sounded remarkably raw compared to the contemporary music charts. Those regular listeners of Radio Luxembourg or the Light Programme must have sat up in shock at the sound of the streets that The Beatles represented. Indeed, some have suggested that "Love Me Do" was such a relatively primitive sound that cultural conservatism alone explains why sales were cautious and the single version of the song peaked only at number 17 in the British chart.

"That harmonica lurching out of the radio, wailing and sneery-snarling. Then the voices, McCartney high, Lennon low, harmony tasting salt and sour, blank-faced yet angry. The song is about love and sex, direct, no messing," *Mojo* writer Phil Sutcliffe said about "Love Me Do." "'I'll always be true' has to be said, but when McCartney solo growls 'Love Me Do' that's the truth. He's not 'experienced,' as in 'mature,' but he's no soppy teenager either. Back then, it sounded like who we wanted to be."

As McCartney astutely says of "Love Me Do" in *The Lyrics*, "I think our image and our energy as the four Beatles were what was potent. 'Love Me Do' wasn't a major hit; it just sort of crept into the charts. We'd been touring the country since summer 1960, so we had a lot of fans around Britain. We had a very fresh sound; that's the sort of thing people notice. And we had a very fresh image. Nobody looked like us. Before not too long, of course, *everybody* looked like us."

"P. S. I Love You"	Recorded: June 6; September 11, 1962	McCartney *(vocals, bass guitar)*	Lennon *(backing vocals, acoustic rhythm guitar)*
(McCartney-Lennon)	UK Release: October 5, 1962	Harrison *(backing vocals, lead guitar)*	Starr *(maracas)* Andy White *(drums)*

In the 2021 six-part miniseries *McCartney 3, 2, 1,* American producer Rick Rubin asked McCartney about his recording career. The topic of family background and resulting psychological outlook came up. The pair spoke of the ways in which the contrasting fortunes of Lennon and McCartney as kids was to color their adulthoods. McCartney described his own upbringing as having "a very nice atmosphere. It's funny, I say to people; I thought everyone had loving families. And everyone was very nice to each other. Of course, later I found that's not true. And some people are very unfortunate. John was very unlucky because his dad left his home when he was three, and John didn't see him until he was famous. And also John's mum got killed. You know, this was an eye-opener to me. It was like, wow. I thought everyone lived like we did."

When Rubin then asked McCartney how their contrasting experiences influenced their different temperaments as grown-ups, even suggesting Lennon had a "chip on his shoulder," McCartney was magnanimous in his response: "That's true. John had a very defensive way . . . which was beautiful. It's how he got through that childhood. While I was much more open and . . . optimistic."

This difference in early life experiences and temperament has been theorized as a factor explaining the rich contrast in the pair's songwriting. Lennon's melodies tend to be "horizontal," in that there is relatively little movement "up and down," while McCartney's are mostly "vertical," with relatively marked movement up and down the melodic scale. Arguably, one of the best exemplary comparisons of this contrast is the two 1967 songs, McCartney's "Penny Lane" and Lennon's "I Am The Walrus." "Penny Lane" has an extraordinarily flexible and vertical melody while the lazy irony of Lennon's "I Am The Walrus" has the reputation of being melodically flat if not monotonous (more on these songs later, naturally.)

McCartney's melodic inclination is already present on "P.S. I Love You." Here, as so very often later, the familial optimism of his youth is expressed in the broad rise and fall of his melodies and bass lines. Written by McCartney back in Hamburg in 1961, the song had fast become a staple of the band's stage act through '62 and '63, apparently being especially popular with female fans. Understandable, then, that the band should select it as one of the four songs they chose to demo at their EMI audition on June 6, 1962—it was taped after the remake of "Love Me Do" the following week.

The look and feel of the recording of "P.S. I Love You" is very much *not* of a rock sensibility. Consistent with McCartney's catholic musical upbringing, "P.S. I Love You" has more of the *cha-cha-cha* of Latin dance music, down to the tempo, beat, and choice of percussion. In 2020,

Cameron Colbeck, a communications manager at Abbey Road Studios who specializes in writing and music history, wrote about the "genius" of Paul McCartney. Colbeck not only identified McCartney as the single most "influential individual" in the then eighty-nine-year history of the Studios, but in addition to the blues and rock 'n' roll that was brought over by travelers and sailors in Liverpool in the '50s, Colbeck was also at pains to point out that McCartney is "a lover of show tunes."

And so we can see how many of the band's early remix songs seem eager to fit existing musical tropes. "P.S. I Love You" sounds very much like the latest contribution to the sub-genre of pop "letter" songs, such as Elvis Presley's "Return To Sender," recorded in March of 1962, and the 1961 song "Please Mr. Postman," recorded by The Marvelettes and covered by The Beatles themselves on their next album, *With The Beatles*. As with some versions of "Love Me Do," the "P.S. I Love You" sessions saw Andy White guesting on the drums in this song, with Ringo relegated to maracas.

"Baby It's You"	Recorded: February 11, 1963	McCartney (*backing vocals, bass guitar*)	Lennon (*vocals, rhythm guitar*)
(David-Williams-Bacharach)	UK Release: March 22, 1963	Harrison (*backing vocals, lead guitar*)	Starr (*drums*) George Martin (*celesta*)

"Baby It's You" was a 1961 recording by The Shirelles, which reached number three on the R&B chart and peaked at number eight on Billboard's Hot 100. The song was written by the celebrated Burt Bacharach along with Mack David, elder brother of Bacharach's better-known lyricist, Hal, and Luther Dixon, credited here as Barney Williams. "Baby It's You" is actually the second Shirelles song The Beatles recorded that day, the other being "Boys," which was also co-written by Luther Dixon. One wonders if the Black girl group from New Jersey were intrigued that two of their songs were being covered on an album by an English band from Liverpool, for heaven's sake, let alone by the same band who were about to go stratospheric.

The truth is that The Beatles had been performing "Baby It's You" as part of their stage act from 1961 until 1963, so its inclusion here is hardly a surprise. In contrast to the far raunchier "Boys," and consistent with the band's eagerness to fit contemporary musical tropes, "Baby It's You" is a rather subversive number, dreamy and gaudy in format, yet lyrically it's subtly vicious. The Shirelles version is totally transformed by the gutsiness of Lennon's raw vocal, though on display once more is the band's ironic approach to remix, especially in the arch "Cheat! Cheat!" refrain.

Indeed, Lennon's Zube-d up voice, which had been deteriorating all day and here cracks pleasingly on the "Don't want nobody, nobody" section, was about to break out into the legendarily rasping "Twist and Shout," in which Lennon can be heard giving it everything he had. But, for now, "Baby It's You" was recorded in three takes, one of which was a false start, and the last take labeled as best. The track was done and dusted nine days later when George Martin recorded himself playing celeste over Harrison's guitar solo.

"Do You Want To Know A Secret"	Recorded: February 11, 1963	McCartney *(backing vocals, bass guitar)*	Lennon *(backing vocals, rhythm guitar)*
(McCartney-Lennon)	UK Release: March 22, 1963	Harrison *(vocal, lead guitar)*	Starr *(drums)*

Another ludic remix composition from the band, this time from Lennon, who, legend has it, taped a demo of "Do You Want To Know A Secret" for an artist under the same management named Billy J. Kramer while sitting on a toilet seat in a Hamburg nightclub. The legend also suggests that, after the tape was done, Lennon flushed the toilet, later claiming the bathroom was the only place quiet enough to record it during those hectic Hamburg nights.

At two seconds shy of two minutes, "Do You Want To Know A Secret" was not taken very seriously. Though by no means without its own charms, the song was originally written by Lennon as a gift to Harrison for his solo-spot on stage. The song was remixed from a number of different sources: the tune is heavily borrowed from "Wishing Well" from Disney's *Snow White and the Seven Dwarfs*; the middle section is a Buddy Holly spoof; and elements of the verse are borrowed from McCartney's arrangement of "Till There Was You." Once more there's a live feel to the recording—the backing track is basic, at best; the production is confined to echoed drumstick-clicks; and errors in Harrison's singing are left in the track's mix.

Nonetheless, "Do You Want To Know A Secret" became a hit for Billy J. Kramer and The Dakotas, who took it to the top of the UK charts in June 1963 (Lennon's lavatory audition clearly paid dividends), while The Beatles' own version made its way to No. 2 in the US in May 1964, by which time, in April 1964, the band had managed a feat that no other band has ever come close to: holding the top *five* spots on the Billboard Hot 100.

"A Taste Of Honey"	Recorded: September 11, 1962	McCartney *(double-tracked lead vocals, bass guitar)*	Lennon *(backing vocals, rhythm guitar)*
(Scott-Marlow)	UK Release: March 22, 1963	Harrison *(backing vocals, lead guitar)*	Starr *(drums)*

Continuing with the theme of McCartney as a lover of showtunes, "A Taste Of Honey" seems to be the sort of sticky sweet ballad to which Paul must have always been drawn but did not write for himself until the later Beatles albums. The song is taken from the film of the same name. Directed by Tony Richardson in 1961, "A Taste Of Honey" was part of the "northern new wave" of British cinema of the early '60s, so its inclusion is very topical, especially as the movie starred Liverpudlian actress Rita Tushingham.

The film's theme, previously recorded as instrumentals, had been in the charts in late 1962. So McCartney took those tracks and, with minor modifications to the chorus lyric, conjured his own version of "A Taste Of Honey" when the band were planning *Please Please Me*. As the song was in the charts at the time, "A Taste Of Honey" almost gate-crashed itself onto the album. Given the otherwise raw and live feel of the rest of the *Please Please Me* album, like "Till There Was You," "A Taste Of Honey" had more to do with McCartney's mainstream fancies than the rest of The Beatles. Nonetheless, and in the nature of collective responsibility, Ringo dutifully deploys his brushes, while George and John *"doo-doo-n-doo"* their way through their chorus responses.

"There's A Place"	Recorded: February 11, 1963	McCartney *(vocals, bass guitar)*	Lennon *(vocals, harmonica, rhythm guitar)*
(McCartney-Lennon)	UK Release: March 22, 1963	Harrison *(backing vocals, lead guitar)*	Starr *(drums)*

Few songs signal the sheer ambition of the early Beatles better than "There's A Place." Written only months before at the McCartney family home, the song was given pride of place at the February 11, 1963 session by being the first band-written track recorded that day. Legend has it that a copy of the soundtrack of *West Side Story* was hanging around the McCartney home. The track "Somewhere" features the lines "There's a place for us/Somewhere a place for us," which was enough to inspire the title of this Beatles track that speaks of a youthful yearning for peace away from the eyes of prying adults. As McCartney explained in his authorized biography, *Many Years From Now*: "In our case, the place was in the mind, rather than round the back of the stairs for a kiss and a cuddle. This was the difference with what we were writing. We were getting a bit more cerebral."

Lennon's lyric, which relates to the singer's ability to overcome his loneliness by retreating into the haven of his mind, is a unilateral declaration of youthful independence—a call to arms, and a statement of self-sufficient defiance which is matched by energetic music of pride and some pain. The fact that the song is a milestone in the development of the new and emergent youth culture is strongly supported by the fact that it shot up to No. 2 in the US singles chart in

Please Please Me (1963)

April 1964. American youth, who were otherwise accustomed to the bland commercialization of daily life, were mesmerized. "There's A Place" shows a glimmer of the beginnings of their more ambitious love songs, ultimately leading to songs like "Help!"

"I got a Japanese bootleg album and spent ages listening to twenty-odd takes of them getting this song wrong over and over and then finally getting it together," Graham Coxon of British band Blur said about "There's A Place."

"There's really something about the difference between John and Paul's voices—John's voice is so dry and rough, Paul's is so effeminate and cutesy.... It seems to me it says a lot about the difference between the two men at that time. They were against the clock, with only hours to record the whole *Please Please Me* album, and you can really hear it here. You can't fake that kind of urgency."

Lennon said in an interview in 1980 that "There's A Place" was his "attempt at a sort of Motown, Black thing," emulating the style and feel of classic Motown hits. The song also evokes the usual Lennon tropes: "In my mind, there's no sorrow . . . It's all in your mind." Perhaps it's this intent that drives the song's ferocity. The track was finally set down after thirteen takes, and it is another rough-house performance, reminding us once more that Lennon's voice, peerless at the best of times, sounds even more passionate and lethal when suffering from a heavy cold.

"Twist And Shout"	Recorded: February 11, 1963	McCartney *(backing vocals, bass guitar)*	Lennon *(vocals, rhythm guitar)*
(Medley-Russell)	UK Release: March 22, 1963	Harrison *(backing vocals, lead guitar)*	Starr *(drums)*

When the clock struck 10:00 p.m., the time when the strict schedule at Abbey Road demanded that the studio officially be shut up for the night, despite the superhuman stamina of the day, the band was still one track short. Not only that, but early next morning, they were expected to make the long trek to the north of England to fulfill a booking in Oldham, Lancashire. They needed to get this done. Not for the last time, George Martin was about to bend the rules for The Beatles. He wanted to sneak in one more session so the album ended with a bang. And what a bang it was.

As McCartney would later recall, "We all retired to the studio canteen for coffee and biscuits, where we and George Martin began an earnest discussion about a suitable number for the last track." *New Musical Express* journalist Alan Smith was reporting on the sessions at the time and later said in a BBC documentary: "We all crowded in there, and I think it was George who said, 'What are we gonna do for the last number?' I said, 'I think I heard you do *La Bamba* on the radio a few weeks ago.' McCartney looked a bit blank, and then he said, 'You mean "Twist and Shout"?! I said, 'Yeah, "Twist and Shout"!'"

The plan was immediately settled.

George Martin knew what to expect. He had been to the Cavern and had witnessed in person the fact that the band's version of "Twist and Shout" always brought down the house.

"John absolutely screamed it," Martin later recalled. "God alone knows what he did to his larynx each time he performed it, because he made a sound rather like tearing flesh. That had to be right on the first take, because I knew perfectly well that if we had to do it a second time it would never be as good."

As the band tuned up one final time that evening in Studio 2, the serious question of whether Lennon could manage arose.

"By this time all their throats were tired and sore," studio engineer Norman Smith recalled. "It was twelve hours since we had started working. John's in particular was almost completely gone, so we really had to get it right first time. John sucked a couple more Zubes, had a bit of a gargle with milk, and away we went."

Stripping off his shirt, Lennon stepped up to the microphone. And so, before we go any further, dear reader; do yourself a favor. Right now. Put this book down for a couple of minutes and listen to John

Lennon singing "Twist and Shout" on the *Please Please Me* album. It is one of the greatest rock vocal performances ever committed to vinyl. Lennon gave out a rock 'n' roll wail which, sixty years later, still evokes feelings of awe.

In 1970, he told *Rolling Stone* that he "couldn't sing the damn thing—I was just screaming. The last song nearly killed me. My voice wasn't the same for a long time after. Every time I swallowed, it was like sandpaper. I was always bitterly ashamed of it, because I could sing it better than that. But now it doesn't bother me. You can hear that I'm just a frantic guy doing his best."

"Twist and Shout" epitomizes the raw, live, and naked feel of the band's first album. Through Lennon's voice, we can feel the passion and dedication that more than makes up for the trembling pitch and occasional vocal cracks, which merely add a flawed and beautiful humanity to the song. The rest of the band backed John up by playing with a similar passion, which was particularly impressive given the grueling day they'd had.

Ringo set upon the drums with added fervor, while Paul and George accompanied the lead vocal with close-cut harmonies and supportive whoops. In McCartney's words, [John] "knew his voice had been going all day, and he could only give it one or two goes and it would just rip it—which it did. You can hear it on the record. But it was a pretty cool performance."

The closing moments of the track also recorded an involuntary and genuinely joyous "Hey!"—McCartney's spontaneous reaction to one of Lennon's finest moments.

Sure, a second attempt was tried. But what was the point? Lennon had created near perfection first time around. As George Martin said, "It was good enough for the record, and it needed that linen-ripping sound." Thus, the album closed, and the "Twist and Shout" track was canned with no overdubs, no edits, and no frills. *Please Please Me* was complete, "live" and unleashed on a world unsuspecting of the mania to follow.

WITH THE BEATLES (1963)

Beatlemania!

"I think I had my first orgasm at a Beatles concert; then again, how would I have known? When you're preteen, prepubescent, and pretty much pre-everything, 'I Want to Hold Your Hand' seems the height of erotic ambition. And that was especially true in 1964, before the sexual revolution and the Internet made that kind of ignorance unimaginable."

—Jeannette Catsoulis, *The Beatles Awaken a New Sensation, New York Times* (2015)

With The Beatles	Released: November 22, 1963	Recorded: July 18, 1963– October 23, 1963	Duration: 33:07
Producer: George Martin	Studio: EMI, London	Label: Parlophone	Tracks: 14

Track Listing

Side One

No.	Title	Lead Vocals	Length
1	"It Won't Be Long"	Lennon	2:13
2	"All I've Got To Do"	Lennon	2:02
3	"All My Loving"	McCartney	2:07
4	"Don't Bother Me"	Harrison	2:28
5	"Little Child"	Lennon with McCartney	1:46
6	"Till There Was You"	McCartney	2:14
7	"Please Mr. Postman"	Lennon	2:34

Side Two

8	"Roll Over Beethoven"	Harrison	2:45
9	"Hold Me Tight"	McCartney	2:32
10	"You Really Got A Hold On Me"	Lennon and Harrison	3:01

(Continued)

With The Beatles (1963)

11	"I Wanna Be Your Man"	Starr	1:59
12	"Devil In Her Heart"	Harrison	2:07
13	"Not A Second Time"	Lennon	2:07
14	"Money (That's What I Want)"	Lennon	2:49

All songs written by Lennon-McCartney except: track 4, written by Harrison; track 6, written by Meredith Wilson; track 7, written by Georgia Dobbins, William Garrett, Brian Holland, and Robert Bateman; track 8, written by Chuck Berry; track 10, written by Smokey Robinson; track 12, written by Richard Drapkin; and track 14, written by Janie Bradford and Berry Gordy.

With The Beatles

From the release of *Please Please Me* in March 1963 to the release of their second album, *With The Beatles*, eight months to the day later in November 1963, Beatlemania had exploded across Britain. *With The Beatles* was the most eagerly anticipated release of the decade to date. The album had incredible advance orders of three hundred thousand and sold another half a million by September 1965, soon making it the first Beatles album to sell a million copies in the UK. The seemingly timeless *With The Beatles*, an album full of exuberant, unbridled joy, is still regarded as one of the band's seminal works by both Beatles fans and critics.

Our focus in this chapter will be how *With The Beatles* reveals just how quickly the band were maturing, both as songwriters and inside the studio. Compared to the headlong rush of the recording of *Please Please Me*, *With The Beatles* was a far more sedate affair, recorded over an extended period of three months from July to October 1963. Nonetheless, it actually took the equivalent of just seven non-consecutive days to complete, plus several editing and mixing sessions. These recording sessions were slotted in around a punishing schedule of numerous concerts, radio and television appearances, and other public engagements. Lennon described *With The Beatles* as the true sound of The Beatles.

Geoff Emerick remarked in his book *Here, There and Everywhere: My Life Recording the Music of the Beatles*: "As I sat in the control room listening to the tracks, I was amazed at how much The Beatles had improved since their debut album, in terms of both their musicianship and their singing. There was a lot more confidence in their individual performances." All this in eight months.

As the band were still bound by the technological limitations of the time, *With The Beatles* was recorded entirely on the same two-track machines used for *Please Please Me*. It was only from "I Want To Hold Your Hand," recorded mid-October 1963, that the group moved on to four-track recording. But the band's hunger for sonic experimentation in their constant search for new sounds meant that they used overdubs and double tracking to a greater extent, as well as a wider use of percussion and keyboard instruments.

While recording "She Loves You" at the start of July 1963, sound engineer Norman Smith had made some technical improvements which would have a profound effect on *With The Beatles*. Geoff Emerick remembers in *Here, There and Everywhere*: "Clearly, [Norman] had been thinking about how he wanted to improve the sound of Beatles records, and on this session he made two significant changes. First, using an electronic device called a 'compressor,' he decided to reduce the dynamic range—the difference between the loudest and softest signals—of the bass and drums independently of each other; in the past, they had been compressed together because they comprised the rhythm section. Second, he specified that a different type of microphone be suspended over the drum kit—the 'overhead' mic, as it was known. The result was a more prominent, driving rhythm sound: both the bass and drums are brighter and more 'present' than in previous Beatles records."

Not only did The Beatles experiment with studio technology, they also experimented with new equipment during the recording sessions for *With The Beatles*. Photographs from the sessions show John and George tying out the Gibson Maestro Fuzz-Tone distortion box, primarily on George's track "Don't Bother Me." Even though the results did not appear on the record, it predates The Rolling Stones' use of

With The Beatles (1963)

the Maestro Fuzz on "Satisfaction" by almost two years; where The Beatles led, the Stones soon followed. The Beatles abandoned the use of the fuzz box in favor of an amplifier tremolo (the Vox AC-30 amplifiers had a built-in tremolo), and this became the group's first use of an electronic guitar effect in the studio. The first of many.

Such was the growing fame of The Beatles at this time that they were drawing attention from the "serious" music critics in the press. *The Times* newspaper's music critic, William Mann, in his essay "What Songs The Beatles Sang," praised Lennon and McCartney as the outstanding English composers of 1963, "Harmonic interest is typical of their quicker songs, too, and one gets the impression that they think simultaneously of harmony and melody, so firmly are the major tonic sevenths and ninths built into their tunes, and the flat submediant key switches, so natural is the Aeolian cadence at the end of 'Not A Second Time' (the chord progression which ends Mahler's 'Song of the Earth')." Lennon was bemused with the critique, "I still don't know what it means at the end, but it made us acceptable to the intellectuals. It worked and we were flattered. I wrote "Not A Second Time" and, really, it was just chords like any other chords. To me, I was writing a Smokey Robinson or something at the time." Indeed Lennon considered that "Aeolian cadences" sounded like "tropical birds"! In response to Mann's essay, *Rolling Stone* suggested "It Won't Be Long" was "the kind of song Bob Dylan had in mind when he wrote that Beatles chords were 'outrageous, just outrageous.'"

During the recording sessions for *With The Beatles*, the band also recorded their next single, "I Want To Hold Your Hand." The track is probably one of the most instantly recognizable songs in the entire history of popular (or any other) music. "I Want To Hold Your Hand" is important technologically speaking too. It was the first Beatles song to be recorded using four-track tape recording technology. The new Studer J-37 four-track machine had just been installed at Abbey Road studios. The four-track technology offered new recording permutations and allowed the band more flexibility in order to carry out studio experiments in their continuous quest for new sounds.

According to Geoff Emerick, again in *Here, There and Everywhere*, the switch to four-track was well-earned, "Apparently the bigwigs at EMI had decided that the band had now earned sufficient monies for the label—many millions of pounds, for sure—to be afforded the same honor as 'serious' musicians, none of whom, I am certain, brought in even a fraction of the income that The Beatles did." The Beatles had come to the session very well prepared with their new song. Emerick remembers, "I wasn't disappointed. From the very first run-through, it was apparent that 'I Want To Hold Your Hand' had 'hit' written all over it, and there was little George Martin could do to improve it. In fact, their first rehearsal sounded very much like the finished version on record, other than some minor changes in tempo and vocal harmony; every one of the takes they did that afternoon sounded confident, professional, and polished. Clearly, a great deal of thought and rehearsal had gone into 'I Want To Hold Your Hand.' I wondered when they'd had time to do so in the midst of their grueling touring schedule."

The Beatles used the new four-track technology to great effect. As Emerick explained in *Here, There and Everywhere*, "The luxury of working in four-track instead of two-track gave Norman a great deal more control over the balance of the instruments. His general way of allocating the four tracks during Beatles sessions was to put drums and bass on one track, combine Lennon's and Harrison's guitars on another, and then put the vocals on a third track. The fourth track was the 'catch-all' track for whatever sweetening George Martin wanted to add—handclaps, harmonica, keyboards, guitar solo, whatever. For 'I Want To Hold Your Hand,' George wanted the double-time handclaps to occur in each verse."

Incidentally, "I Want to Hold Your Hand" is one of only two Beatles tracks to be recorded in German (the other being *"Sie liebt dich"* / "She Loves You.") *"Komm, gib mir deine Hand"* (literally "Come, give me your hand") was recorded on January 29 1964 at EMI's Pathe Marconi Studios in Paris. It was only one of a handful of times that the group had recorded outside of Studio 2 at Abbey Road studios. But The Beatles detested the idea. According to George Martin in the book *The Complete Beatles Recording Sessions*, "Odeon (EMI's West German branch) was adamant. They couldn't sell large quantities of records unless they were sung in German." Martin continued, "I thought that if they were right then we should do it. The Beatles didn't agree, but I persuaded them. Odeon sent over a translator from Cologne to coach the boys although they did know a little German from having played there."

However, the band needed a little encouragement to actually turn up at the studio. As Martin explained in *The Complete Beatles Recording Sessions*, "I fixed the session for late-morning. Norman Smith, myself, and the translator, a chap named Nicolas, all got to the studio on time, but there was no sign of The Beatles. We waited an hour before I telephoned their suite at the George V hotel. Neil Aspinall answered, 'They're in bed, they've decided not to go to the studio.' I went crazy—it was the first time they had refused to do anything for me. 'You tell them they've got to come, otherwise I shall be so angry it isn't true! I'm coming over right now.' So the German and I jumped into a taxi; we got to the hotel and I barged into their suite, to be met by this incredible sight, right out of the Mad Hatter's tea party. Jane Asher, Paul's girlfriend, with her long red hair, was pouring tea from a china pot, and the others were sitting around her like March Hares. They took one look at me and exploded, like in a school room when the headmaster enters. Some dived onto the sofa and hid behind cushions, others dashed behind curtains. 'You are bastards!' I screamed, to which they responded with impish little grins and roguish apologies. Within minutes we were on our way to the studio. They were right, actually. It wasn't necessary for them to record in German, but they weren't graceless, they did a good job."

Track by Track: *With The Beatles*

"It Won't Be Long"	Recorded: July 30, 1963	McCartney *(backing vocals, bass guitar)*	Lennon *(double-tracked vocals, rhythm guitar)*
(Lennon-McCartney)	UK Release: November 22, 1963	Harrison *(backing vocals, lead guitar)*	Starr *(drums)*

"It Won't Be Long" is one of those relatively early Lennon and McCartney songs which was actually a collaboration. Whereas Lennon claimed the song in several interviews, McCartney described the song in the 1990s as no doubt dominated by Lennon, but written together, "John mainly sang it so I expect that it was his original idea, but we both sat down and wrote it together." And what a performance it is again from Lennon; his double-tracked lead vocal starts off unaccompanied with a delivery whose visceral impact just about knocks the listener off their chair.

Even though the track's opening riff is strangely reminiscent of Beethoven's Fifth in its hammering insistence, the song is classic early Beatles. It showcases the band's now-famous hallmarks. The call-and-response yeah-yeahs. The scaling guitar licks. It's also typical of the band's early work in that the song has a stagey ending, somewhat like "She Loves You," which the band had just recorded and was about to be released. When the music of "It Won't Be Long" stops, it allows Lennon the briefest solo vocal flourish before the track ends on the boys doing their barber shop best.

"All I've Got To Do"	Recorded: September 11, 1963	McCartney *(backing vocal, bass guitar)*	Lennon *(lead vocals, rhythm guitar)*
(Lennon-McCartney)	UK Release: November 22, 1963	Harrison *(backing vocals, lead guitar)*	Starr *(drums)*

In "All I've Got To Do," the pervasive American influence on the band is quite apparent. Creatively it's another Lennon song in which he is trying to ape both Smokey Robinson and Arthur Alexander. A restlessly dark and moody ditty, the track is one of three songs on *With The Beatles*, along with "It Won't Be Long" and "Not A Second Time," where Lennon is the chief writer. And it is avowedly written for the American market, as the idea of calling a girl on the phone was unthinkable for an ordinary British lad in 1963. As Lennon is quoted as saying in *The Beatles Off the Record* regarding "No Reply" (from the album *Beatles For Sale*), "I had the image of walking down the street and seeing her silhouetted in the window and not answering the phone, although I have never called a girl on the phone in my life! Because phones weren't part of the English child's life."

Two other features of "All I've Got To Do" are worthy of note. Firstly, with respect to the rest of the band's original songs recorded to this point in 1963, the humming and fade of the outro are both new and unique experiments, signaling the continuing creative trend that they exemplify. Moreover, the hummed outro is also a clever musical motif as it chimes with the subtext of the singer's self-satisfied lyrics. Secondly, according to author Dennis Alstrand in his book *The Beatles and their Revolutionary Bass Player*, "All I've Got To Do" is the first time in rock music in which the bass player plays chords as a vital part of the song.

"All My Loving"	Recorded: July 30, 1963	McCartney *(double-tracked vocals, bass guitar)*	Lennon *(harmony vocals, rhythm guitar)*
(Lennon-McCartney)	UK Release: November 22, 1963	Harrison *(backing vocals, lead guitar)*	Starr *(drums)*

In his 1980 *Playboy* interview, Lennon said of "All My Loving," "[I]t's a damn good piece of work. . . . But I play a pretty mean guitar in back." And in *The Lyrics*, McCartney makes some profound points about the recipe of experimentation that the band brought to bear on their writing and recordings, "It's one of the few songs I've written where the words came first. That almost never happens, I usually have an instrument with me. . . . I made my way to a piano and then somehow found the chords. At that point, it was a straight country-and-western love song. With songwriting, you conceive of it in one genre (because you can't conceive of things in thousands of genres), and you have one way of hearing it. If you get it right, however, you realize it has a certain elasticity; songs can be flexible. And when other members of The Beatles would get into the studio, often that's when that elasticity would kick in."

McCartney explained to Rick Rubin during the TV miniseries *McCartney 3, 2, 1* that Lennon's rhythm guitar part for "All My Loving" was pretty tricky to play, even for just two minutes. He wrote in *The Lyrics*: "The thing that strikes me about the 'All My Loving' recording is John's guitar part; he's playing the chords as triplets. That was a last-minute idea, and it transforms the whole thing, giving it momentum. The song is obviously about someone leaving to go on a trip, and that driving rhythm of John's echoes the feeling of travel and motion. It sounds like a car's wheels on the motorway, which, if you can believe it, had only really become a thing in the UK at the end of the fifties. But, it was often like that when we were recording. One of us would come up with that little magic thing. It allowed the song to become what it needed to be."

"Don't Bother Me"	Recorded: September 12, 1963	McCartney *(bass guitar, claves)*	Lennon *(rhythm guitar, tambourine)*
(Harrison)	UK Release: November 22, 1963	Harrison *(double-tracked vocals, lead guitar)*	Starr *(drums, bongos)*

There are two origin stories for Harrison's "Don't Bother Me." The first comes from Liverpool journalist Bill Harry who claims that he challenged George to write a new song, and that George's reply of "Don't bother me" became the inspiration for the song's title. The second comes from August 1963, when the band had a residency in the English coastal resort town of Bournemouth. Feeling poorly one day, Harrison was prescribed an elixir and bed rest by a local doctor. With little else to occupy his time at the Palace Court Hotel, Harrison began not only writing a song, but also recording himself on a primitive tape recorder. The recording survives and features George working on the bridge of the song (recall that a popular songwriting format is the ABABCB structure, in which the A section is a verse, the B section is a chorus, and the C section is the bridge. In this arrangement, the bridge is the part of the song that connects one chorus to another.) George can also be heard whistling through the song's melody, most likely as the song did not yet have lyrics. In his 1980 autobiography, *I, Me, Mine*, Harrison says the song was "an exercise to see if I *could* write a song. . . . I was sick in bed—maybe that's why it turned out to be 'Don't Bother Me.'

To get a better glimpse of the technical recipe that George Martin used on this track, it's worth listening to the single-track vocal on take thirteen, which acted as the base track upon which the second vocal was eventually overdubbed. You will notice there are no other backing voices, of course, and you'll get a better sense of double-tracking's power: the base track has many notes sung slightly out of tune, which become aurally "invisible" on the final mix. Moreover, the original rhythm track has reverb maxed up on the guitar parts and a cacophonous battery of world-music percussion, which was overdubbed by Martin and played by the other three Beatles along with Harrison's second vocal.

"Little Child"	Recorded: September 11–12 and October 3, 1963	McCartney *(vocals, piano, bass guitar)*	Lennon *(vocals, rhythm guitar, harmonica)*
(Lennon-McCartney)	UK Release: November 22, 1963	Harrison *(lead guitar)*	Starr *(drums)*

Though McCartney describes "Little Child" as a short "album filler," with the melody of the line "I'm so sad and lonely" remixed from the song "Whistle My Love" by British guitarist Elton Hayes, there is a very live feel to this one. Though the track is relatively simple, it's given depth by its radical overdubbing. Lennon's vocal part is double-tracked, with McCartney joining him for brief flashes of harmonic brilliance, while other overdubs include Paul on piano and John on harmonica most of the way through. Another quirky item on Martin's recipe for this track is the appearance of a very live sounding instrumental solo which follows a twelve-bar blues format that doesn't appear in the rest of the song! "Little Child" might be an album filler but it sure taps into that same live and unleashed feel of *Please Please Me*.

"Till There Was You"	Recorded: July 18/30, 1963	McCartney *(vocal, bass guitar)*	Lennon *(acoustic rhythm guitar)*
(Meredith Willson)	UK Release: November 22, 1963	Harrison *(acoustic lead guitar)*	Starr *(bongos)*

We wrote earlier about the influence of "showtunes" on the remix development of McCartney's musical taste. To some extent, the same is true, though less directly, of the evolution of Lennon's musical aesthetic. Thus, the famous pair's relationship with the songwriting that went before is somewhat complex. Ambition dictated that, in order to hit the heights of fame, Lennon and McCartney needed their own writing to be at least as good as the considered standards of old. Both had experienced childhoods steeped in show music; McCartney through his self-taught musician father Jim, and Lennon through his late mother Julia. The influence on the two budding musicians, however, contrasted sharply. McCartney was very fond of show tune kitsch, no doubt reminiscent of large family get-togethers. Lennon was left ambivalent, stranded somewhere between the mawkishness of show music, which he abhorred, and its sometimes-ethereal nature, which he liked.

As the first recording session for *With The Beatles* drew to a close, "Till There Was You" was the last track attempted before a busy schedule took the band elsewhere for the time being. Three takes of the track were attempted, then a further eight a fortnight later, with Martin deciding the sticking point being the harmonic acoustic arrangement which

With The Beatles (1963)

was intricate enough to expose the slightest error. When the band revisited the track, Harrison's now much-practiced guitar solo was tastefully conceived and carried out with considerable poise and great nuance. Indeed, Martin's acoustic arrangement, with its Latin beat and Ringo's bongos, is a far cry from the sticky schmaltz of the original version in *The Music Man* Broadway show. Meanwhile, just to give the cover a good British stamp, Martin condoned McCartney's engagingly English vocal, sung without softening the "t" of "at all" into the transatlantic "d" which usually dictated a singer's microphone technique.

"Please Mr. Postman"	Recorded: July 30, 1963	McCartney *(backing vocals, bass guitar)*	Lennon *(double-tracked vocals, rhythm guitar)*
(Dobbins, Garrett, Holland, and Bateman)	UK Release: November 22, 1963	Harrison *(backing vocal, lead guitar)*	Starr *(drums)*

"Please Mr. Postman" features a fascinating early Beatles use of antiphony. Traditionally, antiphonal music is the type performed by two interacting choirs, often singing alternative phrases. Though the phrase antiphon originates as a short chant in Christian ritual, sung as a refrain, the term antiphony is also used to describe the type of call-and-response style of singing exemplified by the following lyrics:

> *WAIT!*
> *Wait, oh yes, wait a minute, Mister Postman*
> *Wait, wai-ai-ai-ait, Mister Postman*
> *Mister Postman, look and see*
> *(Oh, yeah) Is there a letter in your bag for me?*
> *(Please, please, Mister Postman) I been waiting a long, long time*
> *(Oh, yeah) Since I heard from that girl of mine*

Unlike the original 1961 hit record of "Please Mr. Postman" for The Marvelettes, The Beatles version uses a more impactful and dramatic antiphonal counterpoint between the backing and lead vocals from the get-go of the track. This device would soon become another major motif of the band's original work. Think about how, like their cover of "Please Mr. Postman," the song "Help!" also starts with a vocal antiphony, "HELP!" front and center of the intro.

"Roll Over Beethoven"	Recorded: July 30, 1963	McCartney *(bass guitar, claps)*	Lennon *(rhythm guitar, claps)*
(Berry)	UK Release: November 22, 1963	Harrison *(double-tracked vocals, lead guitar, claps)*	Starr *(drums, claps)*

Historically in rock music, global trumps local. Most British rock and pop stars have traditionally sung in an American accent. It seems to matter not from where in old Britain the singer hails. Nor does it seem to matter how they sound when they speak in everyday life. When the song starts, the local accent usually ends. Instead, some kind of generic American accent normally spouts out. Not readily recognizable in terms of region, but definitely more US than UK. Think about the singing styles of London-born Mick Jagger and Adele, Newcastle-born Sting, or Birmingham-born Ozzy Osbourne; each singer's real-life accent shows strong regional features, but all of them adopted a distinctly American singing style, especially Jagger.

How so? Well, for one thing, linguistically, the process of singing has a kind of neutralizing effect on accents. And for another, there is a social expectation, based on musical history, that rock music will be sung in such a way. It's not so much that singers are trying to sound American as much as they are adopting the default style for their genre.

And yet such accent neutralization is not a given. Think of the illustrative exceptions over the years. Those that deliberately sing in London cockney accents, such as Madness, Lily Allen, and, occasionally, Ray Davies of The Kinks and David Bowie, especially when Bowie is doing his "Anthony Newley voice." Likewise, The Proclaimers have traditionally sung in their local Scottish accents and Cerys Matthews in her local Welsh accent. Indeed, the British punk movement which started in the

With The Beatles (1963)

late 1970s had adopted a British accent as the default style for their subgenre. One only has to think of the "singing" style of Johnny Rotten in the Sex Pistols song "God Save The Queen" where his drawn out, "We mean it, maaaan" is not just a rejection of "hippie" speak, but also British punk's refutation of the default American singing style in rock music.

For a famously British band, The Beatles also defaulted to singing in "American." But on some songs, such as "Roll Over Beethoven" and the track "Do You Want To Know A Secret" on *Please Please Me*, George kept his Liverpudlian voice, far more so than Paul and John. This was probably more down to George's inexperience with singing than any punk-like refutation of the American singing style. After all, as Lennon commented with respect to the writing of "Do You Want To Know A Secret," "I thought it would be a good vehicle for [George] because it only had three notes and he wasn't the best singer in the world." Nonetheless, a more conscious and deliberate adoption of the Scouse accent was to return on later albums, like George's "While My Guitar Gently Weeps," John singing on, for example, "Maggie Mae," and "Polythene Pam," and Paul's singing on "Lovely Rita," such as the way he pronounces "military" (*"mil-uh-tree"*) and the Scouse pronunciation of "book" (which rhymes with "Luke").

"Hold Me Tight"	Recorded: September 11-12, 1963	McCartney *(vocals, bass guitar, claps)*	Lennon *(harmony vocals, rhythm guitar, claps)*
(Lennon-McCartney)	UK Release: November 22, 1963	Harrison *(backing vocals, lead guitar, claps)*	Starr *(drums, claps)*

"Hold Me Tight" is a good example of Beatles experimentation. An uptempo rocker written mostly by McCartney several years earlier, the band had run through thirteen takes of the tune during the *Please Please Me* sessions only to result in the track not making the album at all. Those session notes log a frustrating portrait of false starts, breakdowns, and edit pieces to patch up errors. *With The Beatles* was eventually to accommodate "Hold Me Tight" which, when played very loud with McCartney's bass super-boosted, turns out to be a muscular rocker strongly evocative of the band's early live sound.

"You've Really Got A Hold On Me"	Recorded: July 18 and October 17, 1963	McCartney *(harmony vocal, bass guitar)*	Lennon *(vocal, rhythm guitar)*
(Robinson)	UK Release: November 22, 1963	Harrison *(vocal, lead guitar)*	Starr *(drums)*

Additional contributors: George Martin (piano)

With The Beatles also features the first of their four recordings of the Smokey Robinson song, "You've Really Got A Hold On Me." The song later, in 1998, received a Grammy Hall of Fame award, and was also selected as one of "the Rock and Roll Hall of Fame's 500 Songs that Shaped Rock and Roll." The band recorded the song a number of times for BBC Radio in 1963 (one of these, from July 30, 1963, was included on the *Live at the BBC* collection), a live version recorded in Stockholm in October 1963 (released in 1995 on *Anthology 1*) and once again in 1969, during the *Let It Be* sessions, which meant the song also featured in the eponymous 1970 documentary film.

But the original *With The Beatles* cover of the Smokey song was recorded within eight months of Smokey Robinson and the Miracles having a Top Ten hit with the song in November of 1962. In selecting the song, The Beatles showed their usual excellent taste.

"I Wanna Be Your Man"	Recorded: 11-12, 30 September/October 3, 23 1963	McCartney *(backing vocals, bass guitar)*	Lennon *(backing vocals, rhythm guitar)*
(Lennon-McCartney)	UK Release: November 22, 1963	Harrison *(lead guitar)*	Starr *(double-tracked vocals, drums, maracas)*

Famous for being the first hit for The Rolling Stones in late '63, "I Wanna Be Your Man" has a fascinating history. According to various accounts, either the Stones' manager/producer Andrew Loog

With The Beatles (1963)

Oldham or the Stones themselves ran into Lennon and McCartney on the street as the two were returning from an awards lunch. In the words of Mick Jagger from a 1968 interview, "We knew [The Beatles] by then and we were rehearsing and Andrew brought Paul and John down to the rehearsal. They said they had this tune, they were really hustlers then. I mean the way they used to hustle tunes was great: 'Hey Mick, we've got this great song.' So they played it and we thought it sounded pretty commercial, which is what we were looking for, so we did it like Elmore James or something. I haven't heard it for ages, but it must be pretty freaky 'cause nobody really produced it. It was completely crackers, but it was a hit and sounded great onstage." Meanwhile, and more mischievously, Lennon commented on the song in 1980, "It was a throwaway. The only two versions of the song were Ringo and The Rolling Stones. That shows how much importance we put on it; we weren't going to give them anything great, right?"

"Devil In Her Heart"	Recorded: July 18, 1963	McCartney *(backing vocals, bass guitar)*	Lennon *(backing vocals, rhythm guitar)*
(Drapkin)	UK Release: November 22, 1963	Harrison *(double-tracked vocals, lead guitar)*	Starr *(drums, maracas)*

If "Please Mr. Postman" features an early Beatles use of antiphony, "Devil In Her Heart" is an early Beatles mini soap-opera set to music. Point: the backing singers (John and Paul) sing, "She's got the devil in her heart." Counterpoint: the leading man and lead singer (George) ignores them, "But her eyes they tantalize." John and Paul persist, "She's gonna tear your heart apart." But young George is not only adamant and refuses to take their counsel ("Oh, her lips they really thrill me"), he's also sold on the girl in question, hook, line, and sinking heart ("I'll take my chances/For romance is/So important to me/She'll never hurt me/She won't desert me/She's an angel sent to me"). The track is also notable for Lennon's amusing and incessantly demonstrative warnings in the backing vocal (as we discussed earlier on page 16 in "Misery," the band felt they had the freedom to goof about).

"Not A Second Time"	Recorded: September 11, 1963	McCartney *(bass guitar)*	Lennon *(double-tracked vocals, acoustic guitar)*
(Lennon-McCartney)	UK Release: November 22, 1963	Harrison *(acoustic guitar)*	Starr *(drums)*

Additional contributors: George Martin (piano)

"Not A Second Time" is that rare outing for The Beatles—though John's voice is double-tracked, there are no backing vocals on the song. As we have already seen, a general content analysis of the band's compositions suggests that the word "love" is the most common lyric used in all Beatles songs, being used in 42 percent of 312 compositions. Within the world of popular music in the early 1960s, before the decade had developed its later political momentum, "boy meets girl" was a dominant trope for songwriters.

We all know how "boy meets girl" goes. It's one of the oldest and most basic plots: boy meets girl, boy loses girl, boy regains girl. Details vary, but that's the general gist. Even Shakespeare used it, for heaven's sake. In *Romeo and Juliet*: Romeo meets Juliet, loses her due to insane family infighting that ends up with two people dead and Romeo being banished, then finds Juliet again, but in her tomb, so he kills himself. Spoiler.

There are, naturally, many examples of this motif, in film and fiction, in theatre and in music. And within music, one of the variations on the boy meets girl trope is the love-hate relationship which goes something like this: boy, usually, though sometimes girl, is trapped between their rational side which says "go" while their weaker heart cries "stay, at least for now." One just has to think of songs such as the huge 1966 hit for The Supremes, "You Keep Me Hanging On" or "You Really Got A Hold On Me."

With The Beatles (1963)

These traditional compositions place that "go/stay" emotional tension at center stage of the song's lyrics. However, it is telling the extent to which Lennon was capable of trying to hide away the waveringly weaker side of his emotions behind an apparently cold and tough lyrical exterior as he does in "Not A Second Time."

"Money (That's What I Want)"	Recorded: July 18, 1963	McCartney *(backing vocals, bass guitar)*	Lennon *(vocals, rhythm guitar)*
(Bradford-Gordy)	UK Release: November 22, 1963	Harrison *(backing vocals, lead guitar)*	Starr *(drums)* George Martin *(piano)*

Additional contributors: George Martin (piano)

If *Please Please Me* can finish on the iconic cover of "Twist and Shout," why shouldn't *With The Beatles* draw to a similarly resounding cover climax? Formulaic, maybe, but the Beatles' cover of "Money" all but obliterates the original 1959 recording. This track is pure Beatles and pure early 1960s beat-music personified. The bruising twelve-bar blues with the nihilistic lyric perfectly suited Lennon's early persona.

They were well-practiced on this particular track, too; the song had featured regularly in their live repertoire between 1960 and 1964. As a result, "Money" sounds like history being recorded live. The track is a polaroid of a musical movement then provoking a huge boom in "beat music," arousing thousands upon thousands of youths to form bands of their own. It was good for business, too. The groundswell in new young talent encouraged the opening up of hundreds of new venues to accommodate them. Another revolution in music.

Though there would soon be many pretenders able to ape some diluted version of what The Beatles here display, no band would match the sheer power of these vocal performances. "Money" records for posterity two of the most sublime rock voices at full pelt: McCartney's crazed backing harmony, and Lennon's blazing lead which, even by his own peerless standard, has a rasping ferocity. Though the band were known for their wry view on life, they were quite serious about the go-getting spirit of "Money." As they admitted in interview after interview, The Beatles were working on the expectation that, like earlier pop phenomena, they had perhaps three years of peak earning-time in their immediate futures.

When in August of 1963, John was asked how long The Beatles will last, he replied, "People demand that you think 'how long are you going to last?' Well, you can't say, you know. You can be big-headed and say, yeah, we're gonna last ten years, but, soon as you've said that you think, we're lucky if we last three months."

George's answer was even more modest, "I hope to have enough money to go into a business of my own by the time we, erm, do flop [smiles broadly]. I mean, we don't know, it may be next week, it may be two or three years."

A Hard Day's Night (1964)
The Fab Four Go Global

"The audience [at the first concert in Washington DC in February 1964], despite the various parental presences, was mostly teenage, and very hot. In the seat next to me, a little girl was bouncing up and down and saying, 'Aren't they just great? Aren't they just fabulous?' 'Yes, they are,' I said, somewhat inadequately for her, I suppose. 'Do you like them too, sir?' she asked. 'Yes, I do rather,' I said, all too aware that she couldn't understand what this old man was doing sitting next to her! But perhaps she was put more at ease when the boys played a song like 'I Want to Hold Your Hand,' and everybody in the audience started singing with them, for then Judy and I just found ourselves standing up and screaming along with the rest. That may sound daft, but it was exactly the same screaming that adults do at football matches. And for us especially, in the midst of sixty thousand people who were all enjoying themselves to the full, identifying completely with the people who were performing, people we knew intimately, people with whom we had made all the records and every little bit of music—in that situation it was all too easy to scream, to be swept up in that tremendous current of buoyant happiness and exhilaration."
—George Martin, *All You Need Is Ears* (1994)

"The children of the twenty-first century will be listening to The Beatles."
—Brian Epstein, *Interview with Larry Kane* (Philadelphia, 1964)

A Hard Day's Night (1964)

A Hard Day's Night	Released: July 10, 1964	Recorded: January 29, 1964–June 2, 1964	Duration: 33:07
Producer: George Martin	Studio: EMI, London/Pathé Marconi, Paris	Label: Parlophone	Tracks: 13

Track Listing

Side One

No.	Title	Lead Vocals	Length
1	"A Hard Day's Night"	Lennon with McCartney	2:34
2	"I Should Have Known Better"	Lennon	2:43
3	"If I Fell"	Lennon with McCartney	2:19
4	"I'm Happy Just To Dance With You"	Harrison	1:56
5	"And I Love Her"	McCartney	2:30
6	"Tell Me Why"	Lennon	2:09
7	"Can't Buy Me Love"	McCartney	2:12

Side Two

No.	Title	Lead Vocals	Length
8	"Any Time At All"	Lennon	2:11
9	"I'll Cry Instead"	Lennon	1:44
10	"Things We Said Today"	McCartney	2:35
11	"When I Get Home"	Lennon	2:17
12	"You Can't Do That"	Lennon	2:35
13	"I'll Be Back"	Lennon	2:24

All songs written by Lennon-McCartney

Making Television History

In a remarkable feat unlikely ever to be repeated, by the week ending Saturday April 4, 1964, The Beatles sat at all top five positions of the *Billboard* Hot 100 chart. "Can't Buy Me Love," "Twist And Shout," "She Loves You," "I Want To Hold Your Hand," and "Please Please Me" were proudly placed at chart positions 1 to 5 respectively, and, all told, the band held a full dozen places on the US chart.

How did The Beatles break in such a spectacular way in America? For many in the US, their first introduction to The Beatles was a single live and electrifying performance on the famous *Ed Sullivan Show* on Sunday, February 9, 1964. This spectacle alone ignited Beatlemania. If such an explanation seems inadequate, it's worth remembering that a record-setting 73 million people tuned in that evening. It became one of the seminal moments in American television history, and sixty years later, folks still remember precisely where they were the night The Beatles stepped onto Ed Sullivan's stage.

Few now recall that the band shared the limelight with singing banjoist Tessie O'Shea, the cast of the Broadway production of the celebrated musical *Oliver!*, and the impressionist Frank Gorshin, who would soon play the Riddler in the iconic TV version of *Batman*. Their contributions have been mostly lost to history. For on that night, the esteemed Mr. Sullivan, also of Irish descent, was charged with the task of introducing America to the four-headed monster that had hit the US a mere two days before. And so it was that, for a little while at least, "juvenile delinquents" were peacefully glued to the television. Even mom and dad were rapt, and planet Earth tilted a few degrees off its axis. Or so it seemed.

Prior to The Beatles landing on the *Ed Sullivan Show*, of course, Beatlemania had been brewing in Britain for some time. The first time Scottish promoter Andi Lothian had booked The Beatles, in the glacial January of 1963, barely a dozen folk turned out. But by October 5 in the same year, when Lothian brought them to the Glasgow Odeon, the band had had a number one album and three number one singles. It was as if a tornado had torn into town. As Lothian recalled in *The Guardian* in September 2013, "I saw one of them almost getting to Ringo's drum-kit and then I saw forty drunk bouncers tearing down the aisles. It was like the Relief of Mafeking! It was absolute pandemonium. Girls fainting, screaming, wet seats. The whole hall went into some kind of state, almost like collective hypnotism. I'd never seen anything like it."

Legend has it that a *Radio Scotland* reporter turned to Lothian and gasped.

"For God's sake Andi, what's happening?!"

A Hard Day's Night (1964)

Thinking fast on his feet, the promoter replied, "Don't worry, it's only... Beatlemania."

The term "Beatlemania" allegedly first occurred in a *Daily Mirror* report of the band's London Palladium gig eight days later. But Lothian insists the coinage came from him, via *Radio Scotland*.

Whatever the truth, the spectacle predated the term. Wild accounts of teenage girls screaming and sobbing, fainting off then miraculously chasing the band down Britain's backstreets. Police escorts were soon a necessity. But zeitgeist words like "Beatlemania" have a magical power in the media. And, once lit, the phenomenon sparked a fire in the collective imagination.

But from the band's point of view, how did that electrifying performance on the *Ed Sullivan Show* feel?

As McCartney told Letterman in an interview in 2009, with reference to The Beatles appearances on the *Ed Sullivan Show* in 1964 and 1965, "We were only twenty-two. I had to stand somewhere here, and there was a curtain, and the audience was out there, and we were very new to America; loving it, but a little bit scared. And I had to do 'Yesterday'; my song, on my own. And I'd never done this. I'd always had the band with me. But suddenly they said, you're doing 'Yesterday.' So I said, okay. So, I was standing there, 'Come on now, get it together; it's okay.' And the floor manager, the guy on the curtain, came up to me and said, 'Are you nervous?' I said, 'No?' He said, 'You should be, there's 73 million people watching!'"

The TV rating was a record-breaking 45.3, which means 45.3 percent of households with TV sets were watching. The number meant that 23,240,000 American homes had tuned in, with Sullivan's show grabbing a 60 share, meaning 60 percent of the TVs turned on were tuned in to see The Beatles.

A teenage New Yorker in early 1964, American journalist Nicholas Schaffner later wrote in his 1978 book, *The Beatles Forever*, that the band "more than filled the energy gap" left by the demise of 1950s rock 'n' roll for an audience accustomed to the "vacuous" music that had replaced it. Soon, Brian Wilson of the Beach Boys remarked on The Beatles' sudden popularity, saying that they had "eclipsed ... the whole music world," and by early 1964, none other than Bob Dylan recalled that a definite line was being drawn: "This was something that had never happened before ... I knew they were pointing the direction of where music had to go."

A Hard Day's Night

By July 6, 1964, the biggest band on the planet had made their movie debut with *A Hard Day's Night*. In the film, an American director based in the UK, Dick Lester, used a documentary style of filming to capture the claustrophobia of Beatlemania. In the words of the British Film Institute, "If any single director can encapsulate the popular image of Britain in the Swinging Sixties, then it is probably Richard Lester. With his use of flamboyant cinematic devices and liking for zany humor, he captured the vitality, and sometimes the triviality, of the period more vividly than any other director."

In the film's opening sequence, The Beatles are mobbed at a station as they try to board a train for London to film a televised concert. Under the direction of Dick Lester, the movie hits a marvelous energy level. We feel the hysteria of the fans and sense the excitement of the band, as the on-screen action is intercut with the title song; the first time film titles had done such a thing. The effect was to imply that the songs and the adulation were flip sides of the same coin.

From the get-go, it was clear that *A Hard Day's Night* was a different species from the musicals that had starred Elvis and his emulators. Lester's movie was smart. It, like The Beatles, was irreverent; it didn't take itself too seriously. And it featured that electrifying black-and-white, documentary style that followed the band through the early days of Beatlemania. Finally, it was charged with the characters of The Beatles, whose smart-ass one-liners poked fun at the very process of stardom they were experiencing. Reporter: "Tell me, how did you find

A Hard Day's Night (1964)

America?" Lennon: "Turned left at Greenland." "Are you a mod or a rocker?" Ringo is asked at another presser. "I'm a mocker," he replies. Exactly so.

Four days after the film premiered, The Beatles released their third studio album, *A Hard Day's Night*, with side one containing songs from the soundtrack of the film. The album was largely written by Lennon, and offers the richest harvest of harmonies of any single Beatles collection. The resounding success of both movie and album secured the band's status as a transatlantic, if not global, sensation. *A Hard Day's Night* was the first Beatles album to comprise entirely of self-penned songs. Since preciously few artists wrote any of their own material at the time, this was a considerable coup, placing The Beatles on a pedestal they shared only with Bob Dylan. Moreover, many critics consider this album as marking the start of the band's most productive and exciting period of creativity.

Track by Track: *A Hard Day's Night*

"A Hard Day's Night"	Recorded: April 16, 1964	McCartney *(double-tracked vocals, bass guitar)*	Lennon *(double-tracked vocal, rhythm guitars)*
(Lennon-McCartney)	UK Release: July 10, 1964	Harrison *(lead guitar)*	Starr *(drums, bongos)*

Additional contributors: George Martin (piano)

That glorious opening chord of *A Hard Day's Night* has been much written about. Its dramatic effect is related not just to the pitch, but also to that sudden, crisp attack. For those who like to dwell on the technical, the opening chord, a source of constant speculation over the years, was finally said by Harrison, during an online chat on February 15, 2001, to be an F add-9.

However, according to music theorist Walter Everett, there was technically a lot more to it. Everett claims to have cracked the full chord, saying that it has an "introductory dominant function" thanks to

McCartney playing the D in the bass, while the two Georges, Harrison and Martin (on twelve-string guitar and piano respectively) play F A C G over the bass D. The musical euphony is down to Martin's genius in production, of course, but there are two features that stand out: Ringo's drumming and George's twelve-string. Ringo's drumming style is kept constant throughout. With the brief exception of some cowbell beating during the bridges, his wall to wall hammering on drums and cymbals instead of drum fills and texture changes make all the difference. The usual percussive nuance in different song sections is sacrificed for a monolithic wall of sound that enhances the driving thrust of the track.

A Hard Day's Night, the track *and* the album, establishes the band's characteristic sound of George's Rickenbacker 360 Deluxe electric twelve-string guitar. The difference can be a little subtle for some, but the twelve-string configuration typically yields a brighter, more harmonically textured sound, with a stronger attack, than the six-string configuration. George's twelve-string colors much of this album but is most apparent in the opening and closing attack chords of "A Hard Day's Night," and when it is doubled with electric piano in the solo section it becomes one of the most immediately recognizable sound bites in all of popular music. (One imagines that contestants on the American television music game show, *Name That Tune*, would have received little in the way of reward from identifying George's famous opening chord as belonging to "A Hard Day's Night.")

"I Should Have Known Better"	Recorded: February 25/26, 1964	McCartney *(bass guitar)*	Lennon *(double-tracked vocals, acoustic rhythm guitar, harmonica)*
(Lennon-McCartney)	UK Release: July 10, 1964	Harrison *(lead guitar)*	Starr *(drums)*

In January of 1964, Les Beatles had made a mark on Paris at the Olympia music hall. Their season there ran from January 16 until February 4, but the elitist French press were considerably sniffy about the band.

A Hard Day's Night (1964)

"There are too many hairdressers invited to Parisian galas these days for such badly combed people to have a triumph," snarked a critic at *Le Figaro*.

The audience for the premiere were mostly highly lacquered members of the Parisian nightclub and entertainment life. Ringo gloomily predicted that the house would be full of "old people," not much heat would be generated. Luckily, all was not lost. George had bought *The Freewheelin' Bob Dylan*, Dylan's second studio album, on which eleven of the thirteen songs were original compositions. The Beatles became infatuated with Dylan. And so, "I Should Have Known Better" was born. A Lennon composition, based on an imitation of Bob Dylan's huffing harmonica style, this infectious pop song included words which, as their author later confessed, meant next to nothing, including short trite phrases such as "and I do," "can't you see," and "give me more." So, in the band's happy tradition of goofing about, several of the twenty-two takes it took to lay the track dissolved into helpless laughter.

"If I Fell"	Recorded: February 27, 1964	McCartney *(double-tracked vocals, bass guitar)*	Lennon *(double-tracked vocals, acoustic rhythm guitar)*
(Lennon-McCartney)	UK Release: July 10, 1964	Harrison *(12-string lead guitar)*	Starr *(drums)*

In view of what we said earlier about the relatively horizontal melodies of Lennon, compared to the more vertical tunes of McCartney, "If I Fell" is something of a surprise. Not only is it the most chord-intensive song the band had so far recorded, with the chords changing with nearly every note of the tune, the song's raw and emotional honesty owes much to the fact that the melody ranges over an octave; again, unusual for Lennon. And so, despite McCartney's undeniable credentials as a melodist, it's worth noting that three of The Beatles' most romantic early songs (the close-harmony numbers "If I Fell," along with "This Boy," the B-side to "I Want To Hold Your Hand" and "Yes It Is," the B-side to "Ticket To Ride") were mostly, though not necessarily entirely, penned by Lennon. (The rider here is due to the fact that, although Lennon and McCartney often wrote independently, and many of the band's songs are primarily the work of one or the other, it was rare that a song would be completed without some input from both writers.)

It has been argued that, whereas McCartney had an innate melodic flair, Lennon's melodism was more down to the choice of chords. In Peter Jackson's documentary *The Beatles: Get Back*, one could see that McCartney created his tunes independently, and only later went to guitar or piano to work out the chords. Meanwhile, Lennon's tunes appear to feel their way through their harmonies in the style of an instinctive somnambulist, developing the novel sequences and broken phrasing typical of his style. Indeed, George Martin recounted that Lennon conceived of writing songs as "doing little bits which you then join up."

"I'm Happy Just To Dance With You"	Recorded: March 1, 1964	McCartney *(backing vocals, bass guitar)*	Lennon *(backing vocals, rhythm guitar)*
(Lennon-McCartney)	UK Release: July 10, 1964	Harrison *(lead vocals, lead guitar)*	Starr *(drums, African drum)*

The *Hard Day's Night* movie produced an unexpected star in George Harrison. The film's narrative is stitched together by a sharp script which joins up witty bits of dialogue and laugh-out-loud vignettes, mirroring the way Martin reported Lennon as viewing the process of songwriting. In particular, there's a scene where George is being asked to be a 1964 version of an influencer, minus the #ad hashtag of course. George's opinion is being sought on some "fashionable shirts" of which George doesn't approve, "I wouldn't be seen dead in them," he chimes, "they're dead grotty." And so *"grotty"*—short for "grotesque"—entered the lexicon.

Meanwhile, on the album, George's lone moment in the musical spotlight is his double-tracked lead vocal on "I'm Happy Just To Dance With You." Written by Lennon for Harrison, the repeated notes

A Hard Day's Night (1964)

of the song's verse were included to accommodate George's limited range. Recorded during the band's first Sunday morning session, it comprises a backing track of one of McCartney's lush counter-melody bass-lines, plus vocal overdubs. Finally, for those searching for recording "odds and sods," check out the little squeak or scrape that seems to have eluded the quick pulling down of the faders right after the final chord.

"And I Love Her"	Recorded: February 25–27, 1964	McCartney *(vocals, bass guitar)*	Lennon *(acoustic rhythm guitar)*
(Lennon-McCartney)	UK Release: July 10, 1964	Harrison *(acoustic lead guitar)*	Starr *(bongos, claves)*

As can be heard on *Anthology 1*, "And I Love Her" started its evolution as a full electric line-up, including drums. "And I Love Her" is a starkly beautiful composition. The song, written by McCartney for his then-girlfriend, English actress Jane Asher, was described by Lennon as McCartney's "first 'Yesterday'" and by McCartney himself as "the first ballad I impressed myself with."

However, the version of the song that was to become one of the band's most covered numbers (including versions by Cliff Richard, Esther Phillips, and Kurt Cobain!) features a sparse backing track with McCartney's bass joined by acoustic lead and rhythm guitars, and a percussive support of bongos and claves (small cylindrical wooden blocks.) Paul's vocals are double tracked with no other backing.

The main rising and falling melody is sung by McCartney, with Lennon claiming in an interview with *Playboy* that he had contributed the relatively horizontal middle eight ("A love like ours/Could never die/As long as I/Have you near me") which moves around a single chord.

McCartney credits Harrison with creating the signature guitar riff, which makes "a stunning difference to the song." With its cosmic yet intimate metaphors of luminous stars and mysterious sky, the song also greatly benefits from George Martin's sympathetic use of echo to complete that celestial and spacey feel.

"Tell Me Why"	Recorded: February 27, 1964	McCartney *(harmony vocals, bass guitar)*	Lennon *(lead vocals, rhythm guitar)*
(Lennon-McCartney)	UK Release: July 10, 1964	Harrison *(harmony vocals, lead guitar)*	Starr *(drums)*

Additional contributors: George Martin (piano)

In his 1997 book, *Paul McCartney: Many Years From Now*, Barry Miles quotes McCartney on the topic of "Tell Me Why": "I think a lot of [John's] songs like *Tell Me Why* may have been based in real experiences or affairs John was having, or arguments with Cynthia [John's wife] or whatever, but it never occurred to us until later to put that slant on it all."

Lennon himself said the song was "a Black New York girl-group song" (with John, Paul, and George singing a three-part harmony) and with its doo-wop chord changes, which, together with its walking bass and triplet swing, is further evidence of the band's peerless capacity for rhythmic diversity. The only time "Tell Me Why" was performed in front of a live audience was during the filming of *A Hard Day's Night*. The fictitious gig, part of the "studio performance" sequence of the movie, was shot at the Scala Theatre in London on March 31 1964. The audience consisted of 350 hired actors, one of whom was future drummer of Genesis, Phil Collins, who was a mere thirteen years of age at the time of filming. Keen observers will spot that the song was mimed by the band, with the words lip-synced on stage.

"Can't Buy Me Love"	Recorded: January 29, February 25, 1964	McCartney *(double-tracked vocal, bass guitar)*	Lennon *(acoustic rhythm guitar)*
(Lennon-McCartney)	UK Release: March 20, 1964	Harrison *(double-tracked lead guitar, twelve-string guitar)*	Starr *(drums)*

According to data compiled by the British Official Charts Company, "Can't Buy Me Love" was the fourth biggest selling single in their

A Hard Day's Night (1964)

homeland during the 1960s. The Beatles continued to dominate the top ten:

No.	Single	Artist	Record Label	Year
1	"She Loves You"	The Beatles	Parlophone	1963
2	"I Want To Hold Your Hand"	The Beatles	Parlophone	1963
3	"Tears"	Ken Dodd	Columbia	1965
4	"Can't Buy Me Love"	The Beatles	Parlophone	1964
5	"I Feel Fine"	The Beatles	Parlophone	1964
6	"The Carnival Is Over"	The Seekers	Columbia	1965
7	"We Can Work It Out/Day Tripper"	The Beatles	Parlophone	1965
8	"Release Me"	Engelbert Humperdinck	Decca	1967
9	"It's Now or Never"	Elvis Presley	RCA	1960
10	"Green, Green Grass of Home"	Tom Jones	Decca	1966

Indeed, in the top 60, the band also have "From Me To You" at 18, "Hey Jude" and "Hello, Goodbye" at 20 and 21 respectively, "Help!" at 23, "Get Back" and "All You Need Is Love" at 33 and 34 respectively, "Yellow Submarine/Eleanor Rigby" at 40, "Paperback Writer" at 42, "Ticket To Ride" at 44, "Magical Mystery Tour" at 46, "Penny Lane/Strawberry Fields" and "A Hard Day's Night" at 51 and 52 respectively, and finally "Lady Madonna" at 57. (When the band's contract with EMI Records finally ended in 1976, the label retained the right to release existing songs, so EMI ended up reissuing every Beatles single, plus "Yesterday" for the very first time, so in April 1976 the UK Top 100 contained 23 Beatles singles.)

As a mark of the band's burgeoning popularity, "Can't Buy Me Love" also topped the charts in the US, Canada, Australia, Ireland, New Zealand, South Africa, the Netherlands, France, and Sweden. Written by McCartney while in Paris, "Can't Buy Me Love" was composed on the back of the success of "I Want To Hold Your Hand" having just reached number one in America. In *The Beatles Anthology*,

George Martin said that when he first heard a rough version of the song, he felt that "Can't Buy Me Love" needed tweaking:

"I thought that we really needed a tag for the song's ending, and a tag for the beginning; a kind of intro. So I took the first two lines of the chorus and changed the ending, and said 'Let's just have these lines, and by altering the second phrase we can get back into the verse pretty quickly.'"

And what a difference it made.

The basic track was recorded on January 29, 1964, at EMI's Pathe Marconi Studios in Paris, where the band were still committed to their residency at the Olympia Theatre. At this point, the track still included backing vocals, but after hearing this first take, the band felt that the song didn't need them. And so "Can't Buy Me Love" became the first Beatles single released without their characteristic backing harmonies.

The song is another example of the shape of things to come. The track is musically a fusion of affiliated styles, originally conceived with a rolling backbeat and jazzy blues feel, "Can't Buy Me Love" was transformed in just four takes into an ebullient and world-conquering classic. Moreover, lyrically, we have the alchemical casting of "love and money" platitudes into rock and roll, replacing the traditional "my love" with the asexual "my friend" and helping to define the off-handed etiquette of a *blasé* and unsentimental new age. All in all, a foretelling of future Beatles' experimentation.

And, as McCartney wrote in *The Lyrics*, "It's twelve-bar blues, with a Beatles twist on the chorus, where we bring in a couple of minor chords. Usually, minor chords are used in the verse of a song, and major chords bring a lift and lighten the mood in the chorus. We did it the other way round here. The idea is that all these material possessions are well and good, but money can't buy you what you really need. The irony here is that just before Paris, we'd been in Florida where, if not love, money certainly could buy you a lot of what you wanted. But the premise stands, I think. Money can't buy you a happy

A Hard Day's Night (1964)

family or friends you can trust. Ella Fitzgerald recorded the song later that year too, which was a real honor."

"Any Time At All"	Recorded: June 2, 1964	McCartney *(harmony vocals, bass guitar, piano)*	Lennon *(vocals, acoustic rhythm guitar)*
(Lennon-McCartney)	UK Release: July 10, 1964	Harrison *(lead guitar)*	Starr *(drums)*

When one is thinking about the science that The Beatles used in creating and recording their music, one might at first be tempted to think that their songwriting was formulaic. Especially when one reads the typically disparaging comments of Lennon in his 1980 interviews with *Playboy*, where he described "Any Time At All" as "an effort at writing 'It Won't Be Long.' Same ilk: C to A minor, C to A minor—with me shouting!" And yet every creation and recording is deeply nuanced. Among others, here Paul sings the second "any time at all" in the chorus because John couldn't hit those notes, and George Martin reprises a trick from the "A Hard Day's Night" track by using a piano solo echoed note-for-note by Harrison's lead guitar.

"Any Time At All" is one of those Beatles tracks that tends to get overlooked in favor of the more popular hits of its period. And yet the song still showcases a number of compositional techniques which signpost the continuing development of Lennon's songwriting style, and the repeated use of such techniques is significant in the context of the maturation and evolution of the band's music. As American musicologist Alan Pollack points out, "The walking tenor-line in 'Dear Prudence' is, technically speaking, the same old trick as it is here [in 'Any Time At All'], but look at the difference between the two songs!" Indeed, for that matter, just consider the distance traveled, musically, in the mere five years between "Love Me Do" and "A Day In The Life" and you begin to witness the staggering evolution of The Beatles.

"I'll Cry Instead"	Recorded: June 1, 1964	McCartney *(bass guitar)*	Lennon *(double-tracked vocals, acoustic rhythm guitar)*
(Lennon-McCartney)	UK Release: July 10, 1964	Harrison *(lead guitar)*	Starr *(drums, tambourine)*

Composed entirely by Lennon, "I'll Cry Instead" was initially meant to be the musical accompaniment to the "running and jumping" scene in the *A Hard Day's Night* movie, until director Richard Lester replaced it in the film with "Can't Buy Me Love." Perhaps Lester was intimidated by the rather forlorn nature of "I'll Cry Instead" and so opted for McCartney's exhilarating composition. After all, like "You Can't Do That" and "You've Got To Hide Your Love Away" from the same period, "I'll Cry Instead" deals with the aftermath of a breakup, laced with thoughts of self-pity and revenge with little allusion to any past pleasures. "I'll Cry Instead" is pure darkness and pain.

Unlike the band's other early teenage love songs, on "I'll Cry Instead" Lennon sings that, even now as he "cries" over a lost love, he will return to seek vengeance and break the hearts of girls "around the world," and in so doing punish by proxy any girl who had ever rejected him. Little wonder that musician John Kruth said the song was one of Lennon's "stalker songs." Lennon himself later declared that the lyrics of "I'll Cry Instead" professed his newfangled feelings of frustration of worldwide fame coupled with a loss of personal freedom. Cynthia Lennon sympathized. She later said that Lennon's lyrics were a cry for help: "[I]t reflects the frustration he felt at that time," being the "idol of millions . . . [while] the freedom and fun of the early days had gone." Moreover, McCartney later felt that this song, among others, related to problems in Lennon's marriage.

A Hard Day's Night (1964)

"Things We Said Today"	Recorded: June 2–3, 1964	McCartney *(double-tracked vocals, bass guitar)*	Lennon *(acoustic rhythm guitar, piano)*
(Lennon-McCartney)	UK Release: July 10, 1964	Harrison *(lead guitar)*	Starr *(drums, tambourine)*

According to McCartney in *The Lyrics*, "Things We Said Today" was "written in a boat, on holiday in the Virgin Islands" and "now that we were in The Beatles we could afford a boat holiday," hiring a private yacht with a crew. So the song was written (on a cheap acoustic guitar he'd bought to "keep in practice") below deck in his cabin one afternoon to distract McCartney from his seasickness. As McCartney tells it, "That particular day on the boat, I started with an A minor chord. A minor to E minor to A minor, which gave me a sort of folksy, whimsical world. And then in the middle, on 'Me, I'm just the lucky kind,' it goes to the major and gets hopeful. The thing I always loved and still love about writing a song is that, at the end of two or three hours, I have a newborn baby to show everyone. I want to show it to the world, and the world at that moment was the people on the boat." In a very telling quote, McCartney also highlights the question of limited technology available at the time which had a very interesting influence on McCartney's musical memory. As he explains in *The Lyrics*, once he had composed "Things We Said Today" he "had to remember it, of course, because I didn't write it down." He then continues: "It was all in the head. I have wondered since why it was easy for me to remember these things. When I've used a little cassette recorder or some other recording device, I find it hard to remember songs because I haven't made myself remember them. Looking back, I love the fact that my circumstances were as they were. Years later, as I try to explain why I don't read music or write it down, I blame my Celtic tradition, the bardic tradition. The people I come from trained themselves to rely on their memories." This question of limited technology and musical memory is maybe one of the chief reasons that the band generally wrote such memorable melodies.

"When I Get Home"	Recorded: June 2, 1964	McCartney *(harmony vocals, bass guitar, piano)*	Lennon *(lead vocals, rhythm guitar)*
(Lennon-McCartney)	UK Release: July 10, 1964	Harrison *(harmony vocals, lead guitar)*	Starr *(drums)*

With a full complement of self-penned songs and the absence of covers, *A Hard Day's Night* is notable for the various moods, tempos, and musical textures in its mix, some of which already foreshadow the folk-rock style that was to reach an apogee on *Rubber Soul*. In the case of "When I Get Home," we have a rock song, influenced loosely by the Shirelles, with unusual chord progressions. While its provenance in soul music is clear, its ebullient delivery is no doubt in part due to the fact that the band were finally finishing an album begun over three months earlier. Gone now were the days of *Please Please Me* when an album was recorded in one long twelve-hour session.

"You Can't Do That"	Recorded: February 25, 1964	McCartney *(backing vocal, bass guitar, cowbell)*	Lennon *(lead vocal, lead and rhythm guitar)*
(Lennon-McCartney)	UK Release: March 20, 1964	Harrison *(backing vocal, 12-string lead guitar)*	Starr *(drums, bongos)*

The distinctive chiming sound of "You Can't Do That," which was also the B-side of the "Can't Buy Me Love" single, comes down to the Rickenbacker 360 Deluxe electric twelve-string guitar (only the second one ever made and now worth around $10,000) that was presented to George Harrison in 1964 while in New York for the *Ed Sullivan Show*.

Harrison wrote the song's guitar riff on the intro and outro because, according to Tom Petty, George later recalled, "I was just standing there [in the studio] and thought, 'I've got to do something!'" Inspired by the then relatively unknown American soul singer Wilson Pickett, "You Can't Do That" is rooted in twelve-bar blues, and is another of Lennon's tense and threatening songs of sexual paranoia and hypocrisy.

A Hard Day's Night (1964)

As a result of the darker theme, one would expect the song to be unpopular (Richard Lester again dropped it from the final cut of the movie), but "You Can't Do That" was nonetheless a great and favorite song of the period. Interestingly, American singer-songwriter Harry Nilsson recorded a version of the song in 1967 on which references to eighteen other Beatles tunes are on the mix, mostly in the form of lyric snippets hidden in the multi-layered backing vocals. Nilsson's cover has been credited as the first mashup song.

"I'll Be Back"	Recorded: June 1, 1964	McCartney *(harmony vocals, bass guitar)*	Lennon *(double-tracked vocals, acoustic guitar)*
(Lennon-McCartney)	UK Release: July 10, 1964	Harrison *(backing vocals, 12-string lead guitar)*	Starr *(drums, bongos)*

Another esoteric curiosity, "I'll Be Back" is haunting song boasting a touching lyric and flamenco-like acoustic guitars. It's eccentric structure swings between major and minor keys (allegedly based on the chords of Del Shannon's "Runaway") seems to have two different bridges and no chorus. Before you know it, and after little more than two minutes, the song fades away unexpectedly, half a stanza short.

Deliciously, Chicago-based band The Buckinghams (so named to capitalize on the British invasion of the States in the 1960s) released a version of "I'll Be Back" in 1967 which peaked at number one in the Philippines. And so, The Beatles had gone global. Their influence was to be felt everywhere. On October 2, 1964, the day British musician Sting turned a teenager, The Beatles were filming a TV special called *Shindig!*, having just taken America by storm.

"The Beatles were formative in my upbringing, my education," Sting said. "They came from a very similar background: the industrial towns in England, working class; they wrote their own songs, conquered the world. That was the blueprint for lots of other British kids to try to do the same."

BEATLES FOR SALE (1964)
The Start of Studio Experimentation

"There's priceless history between these covers. When, in a generation or so, a radioactive, cigar-smoking child, picnicking on Saturn, asks you what the Beatle affair was all about, don't try to explain all about the long hair and the screams! Just play them a few tracks from this album and he'll probably understand. The kids of AD 2000 will draw from the music much the same sense of well being and warmth as we do today. For the magic of The Beatles is, I suspect, timeless and ageless. It has broken all frontiers and barriers. It has cut through differences of race, age and class. It is adored by the world."

—Derek Taylor, sleeve notes, *Beatles For Sale* (1964)

Beatles For Sale	Released: December 4, 1964	Recorded: August 11 – October 26, 1964	Duration: 33:42
Producer: George Martin	Studio: EMI, London	Label: Parlophone	Tracks: 14

Track Listing

Side One

No.	Title	Lead Vocals	Length
1	"No Reply"	Lennon	2:15
2	"I'm A Loser"	Lennon	2:30
3	"Baby's In Black"	Lennon and McCartney	2:04
4	"Rock and Roll Music"	Lennon	2:31
5	"I'll Follow The Sun"	McCartney	1:49
6	"Mr. Moonlight"	Lennon	2:38
7	"Kansas City/Hey-Hey-Hey-Hey!"	McCartney	2:38

(Continued)

Beatles For Sale (1964)

Side Two			
8	"Eight Days A Week"	Lennon with McCartney	2:43
9	"Words Of Love"	Lennon/McCartney/Harrison	2:04
10	"Honey Don't"	Starr	2:57
11	"Every Little Thing"	Lennon with McCartney	2:04
12	"I Don't Want To Spoil The Party"	Lennon with McCartney	2:33
13	"What You're Doing"	McCartney	2:30
14	"Everybody's Trying To Be My Baby"	Harrison	2:26

All songs written by Lennon-McCartney except: track 4, written by Chuck Berry; track 6, written by Roy Lee Johnson; track 7, written by Jerry Leiber, Mike Stoller, and Little Richard; track 9, written by Buddy Holly; track 10, written by Carl Perkins; and track 14, also written by Carl Perkins.

The Start of Studio Experimentation

The recording sessions for *A Hard Day's Night* were barely in the can when the band jetted off on an ambitious world tour. It began with gigs in Europe, Hong Kong, and Australia, continued with thirty-two concerts in the US and Canada between August 19 and September 20, and finished with a similarly heavy schedule during their tour of the UK in the fall. During this rather exhausting itinerary, Lennon and McCartney wrote most of the uncharacteristically downbeat songs that form the lion's share of the band's fourth album, *Beatles For Sale*. As George Martin said, "They were rather war-weary during *Beatles For Sale*. One must remember that they'd been battered like mad throughout '64, and much of '63. Success is a wonderful thing, but it is very, very tiring. They were always on the go."

The doleful mood of *Beatles For Sale* was reflected in the album's cover. It pictures the unsmiling, knackered-looking lads in an autumnal setting in London's Hyde Park. Described as the very antithesis of the early-1960s pop star image, the cover carried no band logo or credit, and the LP's title was whispered in tiny type atop the photograph, all of which was a major departure compared with standard album artwork of the day.

The pace of the band's success was relentless. Scheduled for Christmas, writing and recording *Beatles For Sale* began in August, just a week after *A Hard Day's Night* had topped the album charts in the UK and the US. Lacking the space and time to create original material following the band's very hectic schedule, they *only* managed eight out of fourteen original songs. The self-penned songs for the second side of *A Hard Day's Night* had been more contemplative and, along with other cultural and musical influences, this evolution continued to develop the technical range of Lennon and McCartney's compositions.

Track by Track: *Beatles For Sale*

"No Reply"	Recorded: September 30, 1964	McCartney *(harmony vocals, bass guitar, claps)*	Lennon *(double-tracked vocal, acoustic rhythm guitar, claps)*
(Lennon-McCartney)	UK Release: December 4, 1964	Harrison *(acoustic rhythm guitar, claps)*	Starr *(drums, claps)*

Additional contributors: George Martin (piano)

Showcasing the bossa nova beat introduced to the world by American musicians Stan Getz and Charlie Byrd, and recorded in a single evening session, "No Reply" took eight takes. The net result shows the band beginning to master the studio. A bed of acoustic guitars lays down a luxurious tone color. Echo and tracking conjure depth and space. And George Martin uses the piano to render a dark and reverberantly fateful feel to the song. On top of all this is Lennon's double-tracked vocals, which sound hugely powerful in the mix. Then, the tension breaks and the song enters the bridge, "If I were you, I'd realize that I/Love you more than any other guy," with its driving drums and stirring spirit. The band must have been well aware of the track's quality and impact as they chose it as an, admittedly downcast, album opener. It's a promising start; a song whose lyric and musical tone chime dramatically and melancholically. No reply; not now, not ever.

Beatles For Sale (1964)

"I'm A Loser"	Recorded: August 14, 1964	McCartney *(harmony vocals, bass guitar)*	Lennon *(vocals, acoustic rhythm guitar, harmonica)*
(Lennon-McCartney)	UK Release: December 4, 1964	Harrison *(lead guitar)*	Starr *(drums, tambourine)*

Whereas *With The Beatles* began with the upbeat song triplet of "It Won't Be Long," "All I've Got To Do," and "All My Loving," *Beatles For Sale* makes its melancholic mark by starting with three sad songs, "No Reply," "I'm A Loser," and "Baby's In Black;" a novelty for the band and one which turned out to be unique. "I'm A Loser" has a tormented lyric, but one which is characteristically masked by a Lennon smirk. As Lennon was later to confess, the song shilly-shallies, "part of me suspects I'm a loser, and part of me thinks I'm god almighty." So the song boasts a self-deprecating marriage of mood and meaning, one which pleased Lennon so much that the track was considered as a potential single.

"Baby's In Black"	Recorded: August 11, 1964	McCartney *(vocals, bass guitar)*	Lennon *(vocals, acoustic rhythm guitar)*
(Lennon-McCartney)	UK Release: December 4, 1964	Harrison *(lead guitar)*	Starr *(drums)*

By the time of the concerts in Paris in early 1965, Lennon was describing "Baby's In Black," somewhat loosely, as a "waltz." Artistically, the song is derived from a variety of influence. Witness its disparate elements: the bluesy melody sung by Lennon and McCartney simultaneously through the same microphone to get a closer feel to the performance, the choice of chords, the folksy country-and-western vocal arrangement, the wonderful swing of Ringo's drums, and the bass in the final verse which yields a most exotic musical tone. When set in a song which has a refrain, a bridge, and guitar solo sections, it's a most delightful creation.

The band must have been pleased with the track as they kept it in their Beatles live act from late 1964 until their final tour in 1966. In

Paul McCartney: Many Years From Now, McCartney is quoted as saying that the band introduced "Baby's In Black" by saying, "'And now for something different' . . . We used to put that in there, and think, 'Well, they won't know quite what to make of this, but it's cool.'"

"Rock and Roll Music"	Recorded: October 18, 1964	McCartney *(bass guitar)*	Lennon *(vocals, rhythm guitar)*
(Chuck Berry)	UK Release: December 4, 1964	Harrison *(acoustic guitar)*	Starr *(drums)*

Additional contributors: George Martin (piano)

A scintillating cover of the 1957 Chuck Berry classic, "Rock and Roll Music," which even eclipses the original, if truth be told. Chuck's recording is more melodic and laid back compared with this harder-driven interpretation. The Beatles' version is even more impressive when you learn that they pulled off the performance in a single take. In a tumultuous cover which reflects the fact that the song was forever part of their live performances between 1959 and 1966, Lennon's vibrancy is life-affirming, especially after many hours of recording in the studio earlier that day. This is no doubt due to Lennon's desire to do right by Chuck Berry and one of his most iconic songs. So Lennon tears through the track, inspiring George Martin to perform a Jerry Lee Lewis on the studio piano.

"I'll Follow The Sun"	Recorded: October 18, 1964	McCartney *(vocals, acoustic lead guitar)*	Lennon *(harmony vocals, acoustic rhythm guitar)*
(Lennon-McCartney)	UK Release: December 4, 1964	Harrison *(acoustic lead guitar)*	Starr *(percussion)*

In *The Lyrics*, McCartney recalls singing "I remember standing in the living room [of 20 Forthlin Road, Liverpool] with my guitar, singing this song. When you think about it, it's a *Leaving of Liverpool* song. I'm leaving this rainy northern town for someplace where more is happening."

He goes on, "It's an interesting melody, too. I'd been searching for striking new combinations of notes. There's something quite original about it. . . . Even if we were open to influencers, one of the great things about The Beatles was our aversion to repeating ourselves. . . . Why repeat ourselves? Why make the same record twice?"

From *A Hard Day's Night* on, The Beatles experimented with blending blues elements with pop and rock. "I'll Follow The Sun" is an example where they conjure something similar, only with folk tropes and pop. The folk vibe is summoned up by the acoustic mix, but the chord progressions and melody are distinctly non-folk. Also of note here is the enchanting harmony vocal of Lennon: he's backing McCartney's vocal most of way through, though, so subtly the listener is seduced into thinking it's a single voice and is treated to a musical alchemy that is utterly captivating.

"Mr. Moonlight"	Recorded: August 14, October 18, 1964	McCartney *(harmony vocals, bass guitar, Hammond organ)*	Lennon *(vocal, acoustic rhythm guitar)*
(Roy Lee Johnson)	UK Release: December 4, 1964	Harrison *(harmony vocals, lead guitar, Africa drum)*	Starr *(percussion)*

From Lennon's introductory vocal scream to McCartney's gloomy Hammond organ, this song is the one cover that many like to hate. Given The Beatles' tendency to joke around, it's unclear how serious the lads are about this blend of doo-wop and Latin styles. The band had added it to their live repertoire in mid-1962, not long after it had been first recorded by Dr. Feelgood and the Interns. Indeed, the song was often their opening number as it commanded the audience's attention before playing even a single note, as Lennon would have to get it right from *nothing,* starting on *that* note ("MISTER . . . Moonlight!") with no chord to precede it.

"Kansas City/ Hey-Hey-Hey-Hey!"	Recorded: October 18, 1964	McCartney *(lead vocals, bass guitar)*	Lennon *(backing vocals, rhythm guitar)*
(Jerry Leiber, Mike Stoller, and Little Richard)	UK Release: December 4, 1964	Harrison *(backing vocals, lead guitar)*	Starr *(drums)*

Additional contributors: George Martin (piano)

One of The Beatles' best ever covers, and recorded in only two takes, "Kansas City/Hey-Hey-Hey-Hey!" is a Little Richard medley from 1959. The band had supported Richard in late '62 both in Liverpool and Hamburg, had heard him perform the medley live on stage, and had incorporated it into their own act until 1964. The band surpass Richard's original recording with the sassy swing of McCartney's walking bass and George Martin's stonking piano. While rehearsing the song, McCartney is said to have found some sections of the song difficult to sing, and later remembered Lennon telling him, "Come on man, you can do it better than this, get up there!" Little Richard had taught McCartney the "secret" of his signature scream while on tour, so during the recording McCartney stood a way off the mike and belted out the lyrics from the back of his throat.

"Eight Days A Week"	Recorded: October 6/18, 1964	McCartney *(vocals, bass guitar, claps)*	Lennon *(vocals, acoustic rhythm guitar, claps)*
(Lennon-McCartney)	UK Release: December 4, 1964	Harrison *(vocals, lead guitar, claps)*	Starr *(drums, claps)*

According to musicologist Alan Pollack, "'Eight Days A Week' provides a fine object lesson in the Beatles art and science of production values; demonstrating an amazing attention to detail in general, and the use of texture changes to help articulate form." "Eight Days A Week" personifies perfect 1960's pop. It's one of those songs which epitomizes the warm, sunny, and positive sound of the decade. Architecturally, the track features the usual lineup of electric, acoustic,

Beatles For Sale (1964)

and bass guitars, along with Ringo's percussion and a most judicious and ingenious use of hand clapping.

Ostensibly a McCartney composition, with a little help from Lennon, it is John who nonetheless sings this song (maybe the key was better suited to Lennon, or a generous and selfless McCartney just realized the double-tracked Lennon voice sounded ideal for the song). Part written before entering the studio, "Eight Days A Week" was the first song that the band brought into the studio unfinished to work on the arrangement during the session. It was a practice that would later come to typify the band's approach. They worked out a middle eight and a harmonized intro which they experimented with in various ways before dropping it altogether on take six. The song was finished in another seven takes, during which time the melody of the actual title-phrase was boldly altered. The intro and outro sections of the track were added during the album's penultimate session twelve days later.

"Words of Love"	Recorded: October 18, 1964	McCartney *(vocals, bass guitar)*	Lennon *(vocals, rhythm guitar)*
(Buddy Holly)	UK Release: December 4, 1964	Harrison *(lead guitar)*	Starr *(drums, suitcase)*

The band's tribute to Buddy Holly, "Words of Love" was recorded in a couple of takes during the protracted session of October 18, 1964. Buddy's 1958 tour of the UK had been a major influence, both on Britain's later beat boom and on The Beatles themselves. Though the band had not included Holly material in their act since 1962, this homage is rather apt, as Holly's original recording of the song was the first known double-tracked pop vocal, with Buddy harmonizing with himself in Everly Brothers style. Here, Lennon and McCartney are the band's version of Phil and Don, as George's Gretsch Tennessean is dialed up to the treble max, and Ringo simulates the sound of hands clapping off the beat by simply slapping a loosely fitted packing case.

"Honey Don't"	Recorded: October 26, 1964	McCartney *(bass guitar)*	Lennon *(acoustic rhythm guitar)*
(Carl Perkins)	UK Release: December 4, 1964	Harrison *(lead guitar)*	Starr *(vocals, drums)*

The band's cover of Carl Perkins' "Honey Don't," the B-side of the original "Blue Suede Shoes." Taped during the last days recording *Beatles For Sale*, even though the band sound dead on their feet, the next day they were back on their extensive tour of Britain.

"Every Little Thing"	Recorded: September 29/30, 1964	McCartney *(harmony vocals, bass guitar, piano)*	Lennon *(double-tracked vocals, acoustic rhythm guitar)*
(Lennon-McCartney)	UK Release: December 4, 1964	Harrison *(double-tracked lead guitar)*	Starr *(drums, timpani)*

Allegedly, like "Eight Days A Week," "Every Little Thing" is an example of a Lennon-McCartney composition where, unusually, the primary composer, in this case McCartney, gifted the song for the other to take on lead vocal, in this case Lennon.

According to Lucy Hebron in her *Far Out* article, "The Beatles song written by Paul McCartney and sung by John Lennon," McCartney told Barry Miles in the biography *Many Years From Now* that "'Every Little Thing,' like most of the stuff I did, was my attempt at the next single," but Paul passed the vocals onto John to try copying the sound and success of "A Hard Day's Night."

McCartney sings harmony vocal with Lennon on the verses, and high harmony on the chorus. The texture of this track is achieved using strongly strummed acoustic guitar and a heavy bassline which is enhanced by doubling in the low octaves using a piano and some rather exotic punctuation by Ringo banging a timpani drum. George's lead guitar is sparse and understated.

Beatles For Sale (1964)

"I Don't Want To Spoil The Party"	Recorded: September 29, 1964	McCartney *(harmony vocals, bass guitar)*	Lennon *(double-tracked vocal, acoustic rhythm guitar)*
(Lennon-McCartney)	UK Release: December 4, 1964	Harrison *(backing vocals, lead guitar)*	Starr *(drums, tambourine)*

An interesting hybrid in which pure Beatles pop meets country. The song was originally written with Ringo in mind, as Ringo was an aficionado of country songs. McCartney explained in *Many Years From Now*, "Ringo had a great style and great delivery. He had a lot of fans, so we liked to write something for him on each album. 'I Don't Want To Spoil The Party' is quite a nice little song, co-written by John and I. It sounds more like John than me so 80-20 to him, sitting down doing a job. Certain songs were inspirational and certain songs were work, it didn't mean they were any less fun to write, it was just a craft." However, it was Lennon who eventually sang the song, and the presentation is again dominated by an acoustic instrumental texture. The song is a subtle take on the archetypal pop trope of party trauma. Whereas hits like the 1963 Lesley Gore classic, "It's My Party" ("and I'll cry if I want to"), describes her kiss-and-tell woes in agonizing detail, Lennon's more poetic take is scant, pensive, and rather Delphic.

"What You're Doing"	Recorded: September 29/30, October 26, 1964	McCartney *(vocals, bass guitar)*	Lennon *(harmony vocals, acoustic rhythm guitar)*
(Lennon-McCartney)	UK Release: December 4, 1964	Harrison *(harmony vocasl, lead guitar)*	Starr *(drums)*

Additional contributors: George Martin (piano)

McCartney wrote "What You're Doing" while in Atlantic City, explaining that, "It's not that Atlantic City is particularly inspiring, it's just that we happened to have a day off the tour there." While McCartney confessed to *Disc* in November of 1964 that the song was a "filler," it still seems to have played the role of a useful experimental track. With its playful drum intro, its jazzy barrelhouse piano, and its quirky internal rhyming schemes ("doing" is paired with "blue and," and "running" with "fun in"), the track has all the hallmarks of a song evolved in the studio, with the band trying out tricks during the recording sessions. Note also the distortion on the guitar backing and solo; a rare sound in 1964. The track concludes with a lovely piano and bass break, showing that the lads were experimenting with ideas and techniques they would later use to such stunning effect.

"Everybody's Trying To Be My Baby"	Recorded: October 18, 1964	McCartney *(bass guitar)*	Lennon *(acoustic rhythm guitar)*
(Carl Perkins)	UK Release: December 4, 1964	Harrison *(vocals, lead guitar)*	Starr *(drums)*

A cover of the 1958 Carl Perkins song, "Everybody's Trying To Be My Baby" was an opportunity for Harrison to take over the mic. George was arguably the band's biggest Perkins fan; during their first tour of Scotland in 1960, as the backing band for Johnny Gentle, the lads decided to adopt pseudonyms, with George becoming briefly known as Carl Harrison. According to Alan Pollack, "The Beatles opt for a thick, muddy sounding instrumental backing that makes Perkins' original look primitive and homespun in comparison."

HELP! (1965)
Riding So High

"I was bored on 9th October 1940, when, I believe, the Nasties were still booming us, led by Madalf Heatlump, who only had one. Anyway, they didn't get me. I attended to varicose schools in Liddypool, and still didn't pass, much to my auntie's supplies. As a member of the most publified Beatles, my P, G, and R's records might seem funnier to some of you than this book, but, as far as I'm conceived, this correction of short writty is the best laff I've ever ready. God help and breed you all."

—John Lennon, *In His Own Write* (1964)

Help!	Released: August 6, 1965	Recorded: February 15, 1965–June 17, 1965	Duration: 33:44
Producer: George Martin	Studio: EMI, London	Label: Parlophone	Tracks: 14

Track Listing

Side One

No.	Title	Lead Vocals	Length
1	"Help!"	Lennon	2:18
2	"The Night Before"	McCartney	2:34
3	"You've Got To Hide Your Love Away"	Lennon	2:09
4	"I Need You"	Harrison	2:28
5	"Another Girl"	McCartney	2:05
6	"You're Going To Lose That Girl"	Lennon	2:18
7	"Ticket To Ride"	Lennon	3:09

Side Two

No.	Title	Lead Vocals	Length
8	"Act Naturally"	Starr	2:30
9	"It's Only Love"	Lennon	1:56
10	"You Like Me Too Much"	Harrison	2:36

(Continued)

11	"Tell Me What You See"	McCartney	2:37
12	"I've Just Seen A Face"	McCartney	2:05
13	"Yesterday"	McCartney	2:05
14	"Dizzy Miss Lizzy"	Lennon	2:54

All songs written by Lennon-McCartney except: track 4, written by Harrison; track 8, written by Johnny Russell and Voni Morrison; track 10, written by Harrison; and track 14, written by Larry Williams.

Track by Track: *Help!*

| "Help!" | Recorded: April 13, 1965 | McCartney *(backing vocals, bass guitar)* | Lennon *(double-tracked vocals, acoustic rhythm guitar)* |
| (Lennon-McCartney) | UK Release: July 23, 1965 | Harrison *(backing vocals, lead guitar)* | Starr *(drums, tambourine)* |

Thanks to the commercial success of the *Hard Day's Night* movie, director Richard Lester was given a bigger budget for the band's second film, which eventually became *Help!* Rather bizarrely, the movie's plot revolves around the band trying to protect Ringo from a creepy eastern cult and a couple of mad scientists. It seems they're all intent on stealing a sacrificial ring sent to Ringo by a Beatles fan. Sure. But, a couple of months into production, the movie was still harboring under the working title of *Eight Arms To Hold You*. Unsurprisingly, neither Lennon nor McCartney were interested in writing a song about an octopod, and Ringo is at this point a good four years away from coming up with "Octopus's Garden." According to Lennon's cousin Stanley Parkes, "Help!" was written after Lennon "came in from the studio one night. 'God,' he said, 'they've changed the title of the film: it's going to be called *Help!* now. So I've had to write a new song with the title called 'Help!'"

The song started as a ballad but was spruced up in the recording studio to create another huge hit. And yet this transformation of the track doesn't diminish its emotional impact. In 1980, Lennon remembered the song as a cry for help from the despair of what he called his "fat Elvis" period. Lennon was exhausted, body and mind, by the recent years of perpetual touring. Moreover, as a working-class lad, he was isolated and alienated in his gargantuan mansion in the jerkwater stockbroker belt of west London.

His marriage was fatally corroded by the debauched carnival of groupies of touring, and he felt unsatisfied by his marriage and his wife's disapproval of his drug use. Lennon had begun using marijuana since late 1964, and around the time of the recording of the "Help!" track, along with George and their wives, he'd had his first experience of the hallucinogenic drug Lysergic acid diethylamide (LSD) when a dentist "friend" spiked their after-dinner coffee. The heightened perception inspired by LSD meant that both musicians became increasingly introspective in their subsequent work with the band.

Musically, the song is so characteristically Lennon that its verse is comprised of one repeated note tailing off into a howl. The track also has a complex backing-vocal, with a pattern that foreshadows rather than responds to the melody. Moreover, the hooks of Ringo's additional tambourine and George's overdubbed guitar licks, which can be heard descending at the end of each chorus, are perfect examples of the band's science and art of painting color into every corner of their compositions. The public adored it. It went to number one on both sides of the Atlantic in the summer of 1965, became the second best-selling single of 1965, and was nominated in four categories at the 1966 Grammy Awards.

"The Night Before"	Recorded: February 17, 1965	McCartney *(double-tracked vocals, bass guitar)*	Lennon *(harmony vocals, electric piano)*
(Lennon-McCartney)	UK Release: August 6, 1965	Harrison *(harmony vocals, lead guitar)*	Starr *(drums, tambourine)*

"The Night Before" is mostly a McCartney composition which features a robust bluesy accent, blended pleasingly by the primarily pop style that underpins it. In this very strong sense, the track represents an exemplar of the way in which the band were developing

Help! (1965)

in mid-career: on a cusp and moving towards some kind of synthesis between the twelve-bar blues of their past and the more progressive passion pop of their future. Biographer Jonathan Gould, author of *Can't Buy Me Love: The Beatles, Britain, and America*, claims the song prefigures the "increasingly complex and conversational" vocal patterns of the band's later songs. It's also the first Beatles song to feature electric piano.

"You've Got To Hide Your Love Away"	Recorded: February 18, 1965	McCartney *(acoustic rhythm guitar)*	Lennon *(vocals, acoustic rhythm guitar)*
(Lennon-McCartney)	UK Release: August 6, 1965	Harrison *(acoustic lead guitar)*	Starr *(tambourine)*

Additional contributors: Johnnie Scott (tenor/alto flutes)

The first Lennon-led vocal for two years not to be double-tracked, "You've Got To Hide Your Love Away" is commonly seen as "basically John doing Dylan," as McCartney put it. Indeed, Lennon himself later said, "That's me in my Dylan period again. I am like a chameleon, influenced by whatever is going on. If Elvis can do it, I can do it. If the Everly Brothers can do it, me and Paul can. Same with Dylan."

Fast-tracked in a single afternoon, with an acoustic backing, no harmony voices, and only light percussion, a melodious flute replaces the trademark jarring harmonica championed by Dylan. "You've Got To Hide Your Love Away" is lyrically cryptic. What kind of love is it that Lennon feels he has to hide away? Perhaps, as an adored Beatle, he is referring to the fact he had had to keep his marriage a secret from the fans. Or maybe (as with "Help!") he is singing about hiding his true "loving" persona from public view, and expressing subsequent feelings of loneliness and paranoia. During the writing of the song, Lennon had sung "two-foot small" (instead of "two-foot tall") to rhyme with "wall" in the first verse, when he auditioned the song to McCartney. But, typically of Lennon, he liked the mistake so much they kept it in.

"I Need You"	Recorded: February 15/16, 1965	McCartney *(harmony vocals, bass guitar)*	Lennon *(harmony vocals, acoustic rhythm guitar)*
(Harrison)	UK Release: August 6, 1965	Harrison *(double-tracked vocals, lead guitar)*	Starr *(drums, cowbell)*

Little wonder Harrison hadn't written a song since the band recorded his first composition, "Don't Bother Me" for *With The Beatles* in 1963, when you consider that George Martin had publicly attributed Harrison's lack of productivity down to the fact that "none of [them] had liked something he had written!"

Nonetheless, Harrison had soldiered on and written "I Need You" for his girlfriend Pattie Boyd (she of "Layla" and Eric Clapton fame), whom Harrison had met while the band were filming *A Hard Day's Night*. The song features the band's first use of a guitar volume pedal. This tone-altering device was a precursor to the wah-wah pedal, and which explains the song's distinguishing lead guitar cadences that act as a hook throughout the track.

"Another Girl"	Recorded: February 15/16, 1965	McCartney *(double-tracked vocals, bass guitar, lead guitar)*	Lennon *(harmony vocals, acoustic rhythm guitar)*
(Lennon-McCartney)	UK Release: August 6, 1965	Harrison *(harmony vocals, electric rhythm guitar)*	Starr *(drums)*

"Another Girl" is a McCartney composition which has an uncharacteristically smug and even cruel lyric ("She's sweeter than all the girls and I've met quite a few"). Supposedly about McCartney's relationship with then-girlfriend Jane Asher, McCartney had allegedly kept a secret flat in London which he used for trysts with various "another" girls.

The sentiment is echoed in the *Help!* movie, where the song is performed while the band are on a coral reef island in the Bahamas, and McCartney is holding a bikini-clad woman, instead of his bass, and is pretending to "play" her. The track is also notable for being one of a handful of songs on which McCartney has overdubbed his own lead guitar part.

"You're Going To Lose That Girl"	Recorded: February 19, 1965	McCartney *(backing vocals, bass guitar, piano)*	Lennon *(double-tracked vocals, acoustic rhythm guitar)*
(Lennon-McCartney)	UK Release: August 6, 1965	Harrison *(backing vocals, lead guitar)*	Starr *(drums, bongos)*

Another vocal tour de force from Lennon, his singing on "You're Going To Lose That Girl" seems to inspire the rest of the band into one of the most powerful performances on the album: from Ringo's fevered showing on the bongos to Paul and George's gutsy reactive backing vocals. The song is archetypal early Lennon, with lyrics that warn an anonymous cuckold of Lennon's predatory intent towards his woman.

With this track, George Martin has achieved one of the best vocal sounds on a Beatles record of this period, while Alan Pollack suggests that the track exemplifies "the Beatles style and sound," explaining that it doesn't matter that their musical "tricks are neither unique to this song nor were necessarily invented by the Beatles themselves," but that "it is the freedom and liberality with which such tricks are deployed throughout the Beatles songbook that stands out dramatically against the . . . pop music of the period."

"Ticket To Ride"	Recorded: February 15, 1965	McCartney *(harmony vocals, bass guitar, lead guitar)*	Lennon *(double-tracked vocals, rhythm guitar)*
(Lennon-McCartney)	UK Release: April 9, 1965	Harrison *(harmony vocals, rhythm guitar)*	Starr *(drums, tambourine, claps)*

Unlike the cute and catchy hits of '63 and '64, the band turn a dark corner for their early '65 hit single, "Ticket To Ride." Lennon declared "Ticket To Ride" one of the earliest heavy metal records ever made, with the track's weighty sound possibly influenced by John and George's recent brush with LSD. Lennon later said that the dazzling and near-hysterical experience of the powerful hallucinogen had stunned him, unveiling a mode of perception which marijuana had merely suggested.

Complex and introspective, "Ticket To Ride" is a driving rock song hinting at the shady underbelly of sexual relations. The sheer gravitas of the track is felt through the texture of high amplification of the jangling electric guitars and Ringo's pounding rhythm on the tom-toms. Once more, it is the result of The Beatles and George Martin trying to aurally paint the impact of the band live.

The track is extraordinary for its time, especially when compared with contemporary hits like The Beach Boys' vacuous "Help Me, Rhonda," and the stagey emotion in the songs of Gene Pitney. It should be noted that there was also a vibe of heightened electric energy in British pop at the time, with songs such as "You Really Got Me" by The Kinks, "Baby Please Don't Go" by Van Morrison's band Them, and "It's All Over Now" by The Rolling Stones. But even in such exalted company, the deliberately oppressive and mesmerizing intense depth of "Ticket To Ride" stands alone. A rubicon had been crossed—one which would eventually lead to the likes of Lennon's "Tomorrow Never Knows."

"Act Naturally"	Recorded: June 17, 1965	McCartney *(harmony vocals, bass guitar)*	Lennon *(acoustic rhythm guitar)*
(Morrison-Russell)	UK Release: August 6, 1965	Harrison *(lead guitar)*	Starr *(vocals, drums, sticks)*

Little else to say about this album filler other than it is the penultimate song of the two dozen non-originals that The Beatles recorded in the studio. "Act Naturally" is a country-and-western cover. It proved a suitable vehicle for Ringo's amiable but limited vocals.

Help! (1965)

"It's Only Love"	Recorded: June 15, 1965	McCartney *(bass guitar)*	Lennon *(double-tracked vocals, acoustic rhythm guitar)*
(Lennon-McCartney)	UK Release: August 6, 1965	Harrison *(lead guitar)*	Starr *(drums, tambourine)*

Developed with a jokey working title of "That's A Nice Hat, It's Only Love" is distinguished by Lennon confessing, "I always thought it was a lousy song. The lyrics were abysmal. I always hated that song," while McCartney was more forgiving, "Sometimes we didn't fight it if the lyric came out rather bland on some of those filler songs like 'It's Only Love.' If a lyric was really bad we'd edit it, but we weren't that fussy about it, because it's only a rock 'n' roll song. I mean, this is not literature."

Lennon's attitude to the song can be heard on the track itself in his snickering delivery of the line, "just the sight of you makes night-time bright . . . very bright." Having said all this, "It's Only Love" has a beautifully lilting melody. Moreover, when you compare the finished product of the varied sound production on Lennon's vocals and Harrison's three electric guitar parts to the rough take on *Anthology 2*, it becomes clear that even on so-called fillers the band is becoming increasingly experimental in their technique.

"You Like Me Too Much"	Recorded: February 17, 1965	McCartney *(backing vocals, bass guitar, piano)*	Lennon *(acoustic rhythm guitar, electric piano)*
(Harrison)	UK Release: August 6, 1965	Harrison *(double-tracked vocals, lead guitar)*	Starr *(drums, tambourine)*

George Harrison's "You Like Me Too Much" boasts an introduction that uses three piano parts: Lennon on a Hohner Pianet electric piano, and McCartney and George Martin playing two contrasting parts on separate ends of the same Steinway grand piano. Originally recorded for inclusion in the *Help!* movie, the track was later demoted to the album's second, non-soundtrack, side.

"Tell Me What You See"	Recorded: February 18, 1965	McCartney *(vocals, bass guitar, electric piano)*	Lennon *(harmony vocals, rhythm guitar)*
(Lennon-McCartney)	UK Release: August 6, 1965	Harrison *(lead guitar)*	Starr *(drums, tambourine, claves, guiro)*

Another song written for *Help!* but rejected by Richard Lester and relegated to the status of filler track on the album's second side, McCartney's "Tell Me What You See" was described by the man himself as, "Not awfully memorable. Not one of the better songs but they did a job, they were very handy for albums or B-sides. You need those kind of sides."

Though the band spent much of the previous year leaning towards bluesy and later folksy directions, "Tell Me What You See" was a return to McCartney's longstanding love for the Latin flavor, showcasing once more the band's drive and determination to explore new directions.

"I've Just Seen A Face"	Recorded: June 14, 1965	McCartney *(vocals, acoustic guitar)*	Lennon *(acoustic guitar)*
(Lennon-McCartney)	UK Release: August 6, 1965	Harrison *(acoustic lead guitar)*	Starr *(brushed snare, maracas)*

McCartney's delightful acoustic number, "I've Just Seen A Face," is an up-tempo, liminal country and western composition about which McCartney confessed to being "quite pleased with." He felt that, "the lyric works: it keeps dragging you forward, it keeps pulling you to the next line, there's an insistent quality to it that I liked."

Due to McCartney's Auntie Gin liking this song, the band gave the track the working title of "Auntie Gin's Theme." Only later did McCartney transform the song by adding the tumbling love-at-first-sight lyric which, together with the composition's breathless lack of space, suits the song to a tee.

Help! (1965)

Recorded in a mere half dozen takes, and without further embellishment, the song's creative urgency appealed to Capitol's A&R department so strongly that they tore it off the US version of the *Help!* album and made it the opening track of the US version of *Rubber Soul*. Had the song been ready earlier, there's little doubt Richard Lester would have used it in his movie.

"Yesterday"	Recorded: June 14/17, 1965	McCartney *(vocals, acoustic guitar)*	
(Lennon-McCartney)	UK Release: August 6, 1965		

Additional contributors: Kenneth Essex (viola), Francisco Gabarro (cello), and Tony Gilbert/ Sidney Sax (violins)

Ensconced in his cosy relationship with girlfriend Jane Asher, McCartney was falling somewhat behind his writing partner. Lennon had penned the band's last four singles, as well as co-written and sung "Eight Days A Week," an additional single issued only in the US. On the Beatles' albums, too, Lennon's material had more variety and gravitas. The same was true, thus far and save for "I've Just Seen A Face," with *Help!* But things were about to change.

Enter "Yesterday," one of McCartney's most famous compositions, and holder of the *Guinness Book of World Records* title as the most covered song in history. "Yesterday" is a famous enough song to render this rumor true: George Martin claims that McCartney had begun composing "Yesterday" during a residence at Paris's George V hotel in January 1964. In other words, during a time when the band were composing for *A Hard Day's Night*, a full two albums before *Help!* If the rumor is true, then the fact that this world-renowned song was left off two Beatles albums might be explained by McCartney being convinced he'd inadvertently "borrowed" the tune from a previous generation: "Is this by someone else, or did I write it?" he'd ask. As McCartney wrote in *The Lyrics*, "Somewhere in a dream, I heard this tune. When I woke up, I thought, *I love that tune. What is it? Is it Fred Astaire? Is it Cole Porter? What is it?"*

To help solidify the tune in his memory, McCartney had used dummy lyrics, a rare act on his part, "scrambled eggs, oh my baby, how I love your legs, scrambled eggs." Once the lyrics had been transformed into "Yesterday," according to McCartney, in the back of a car while driving through Portugal's Algarve region, the band later decided it was best done as a solo McCartney track. So, McCartney "decided to give it a go" and George Martin proposed putting a string quartet on the track,

"Let's just try it. If you don't like it, we can take it off." Martin said.

Using Bach as a reference point and over cups of tea, naturally, the pair came up with a compromise, with McCartney explaining in *The Lyrics*: "Trying to keep things modern, I wanted to add in a few notes that Bach wouldn't have thought to use, so we added in the flattened seventh, which is also known as a 'blue note.' So it has a quite distinctive arrangement."

The recording of the track could have been more radical still. McCartney had considered using the BBC Radiophonic Workshop, one of the sound effects units of the BBC, which was created in 1958 to produce incidental sounds and new music for radio and television. Perhaps the likes of *Doctor Who* had given McCartney the idea to create a futuristic electronic version of "Yesterday":

"It occurred to me to have the BBC Radiophonic Workshop do the backing track to it and me just sing over an electronic quartet. I went down to see them. . . . The woman who ran it was very nice and they had a little shed at the bottom of the garden where most of the work was done. I said, 'I'm into this sort of stuff.' I'd heard a lot about the BBC Radiophonic Workshop, we'd all heard a lot about it. It would have been very interesting to do, but I never followed it up."

Nonetheless, the rest is history. "Yesterday" remains popular today with well over two thousand cover versions. In a 1999 BBC Radio 2 poll of music experts and listeners, the song was voted the best song of the twentieth century and was also voted the number one pop song of all time by MTV and *Rolling Stone* magazine the following year. The US organization Broadcast Music Incorporated have estimated that that "Yesterday" was performed over seven million times in the twentieth century. McCartney had clearly gone some way in catching up with his writing partner.

"Dizzy Miss Lizzy"	Recorded: May 10, 1965	McCartney *(bass guitar, electric piano)*	Lennon *(vocals, rhythm guitar)*
(Williams)	UK Release: August 6, 1965	Harrison *(double-tracked lead guitar)*	Starr *(drums, cowbell)*

A part of the band's live act since the early days, "Dizzy Miss Lizzy" was originally written and recorded in 1958 by American R&B singer and songwriter Larry Williams (he of the hit "Bony Moronie"). Appearing as the album's rock 'n' roll finale to follow McCartney's melancholic "Yesterday," "Dizzy Miss Lizzy" is a twelve-bar blues cover with liberal amounts of near hysterical "ows" and "whoas," which are absent on the original.

Alan Pollack makes the interesting point that, in the mid 1960s, when a White singer was performing the works of a Black artist and was conscious that their cover might suffer from some kind of soul deficit, the White artist would "resort to raspy shouting in order to hit the mark."

"This is not to say that such Beatles' covers are entirely without either merit or success," Pollack says. "But I'd dare say that on some level they sound a bit more parodistic and less interpretive than intended."

RUBBER SOUL (1965)
The Album as Art Form

"I think Rubber Soul *was the first of the albums that presented a new Beatles to the world. Up to this point, we had been making albums that were rather like a collection of their singles and now we really were beginning to think about albums as a bit of art in their own right. We were thinking about the album as an entity of its own and* Rubber Soul *was the first one to emerge in this way."*
—George Martin, *All You Need Is Ears* (2021)

Rubber Soul	Released: December 3, 1965	Recorded: October 12, 1965 - November 11, 1965	Duration: 34:59
Producer: George Martin	Studio: EMI, London	Label: Parlophone	Tracks: 14

Track Listing

Side One

No.	Title	Lead Vocals	Length
1	"Drive My Car"	McCartney with Lennon	2:25
2	"Norwegian Wood (This Bird Has Flown)"	Lennon	2:05
3	"You Won't See Me"	McCartney	3:18
4	"Nowhere Man"	Lennon	2:40
5	"Think For Yourself"	Harrison	2:16
6	"The Word"	Lennon	2:41
7	"Michelle"	McCartney	2:40

Side Two

8	"What Goes On"	Starr	2:30
9	"Girl"	Lennon	1:56
10	"I'm Looking Through You"	McCartney	2:36

(Continued)

Rubber Soul (1965)

11	"In My Life"	Lennon	2:37
12	"Wait"	Lennon and McCartney	2:05
13	"If I Needed Someone"	Harrison	2:05
14	"Run For Your Life"	Lennon	2:54

All songs written by Lennon-McCartney except tracks 5 and 13 written by Harrison

The Album as Art Form

Rubber Soul was a significant milestone for The Beatles, and saw the band come of age. Not only did the recording help to finally break the chains of their mop-top era, but it also signaled the transformation of the Fab Four from a singles band to an album band and helped pave the way for the incredible experimentation that would follow.

The band had finished their US tour on August 31, 1965, and the recording of *Rubber Soul* began in mid-October, so they had just six weeks to write *Rubber Soul*. But at least they were finally able to record an album free of media and concert commitments. With this album, the band truly recovered their sense of purpose, a sense that had been subdued during the hectic work schedules and indulgence in cannabis that had hampered their work on *Beatles For Sale* and *Help!*

Due to their growing experimentation with LSD, which highlighted the contrast between their personal journeys and their overly clean and polished public image, the lads realized the separation of their professional and personal lives was a false dichotomy. And so, the deliberate experimentation of *Rubber Soul* started to define not only a new style, but also the second half of their career.

Track by Track: *Rubber Soul*

"Drive My Car"	Recorded: October 13, 1965	McCartney *(vocals, bass guitar, piano)*	Lennon *(vocals)*
(Lennon-McCartney)	UK Release: December 3, 1965	Harrison *(harmony vocals, lead guitars)*	Starr *(drums, cowbell, tambourine)*

McCartney writes in *The Lyrics* about "Drive My Car":

> I know there's a theory that rock and roll couldn't have existed without the guitars of Leo Fender, but it probably couldn't have existed without Henry Ford either. I'm thinking of the relationship between the motorcar and what happens in the back seat. We know that people shagged before the motorcar, but the motorcar gave the erotic a whole new lease on life. Think of Chuck Berry "riding along in my automobile." Chuck is one of America's great poets.

As discussed in earlier sections, the band's sense of humor, and their healthy tendency to joke around in the studio, was a constant and important theme throughout their career. With "Drive My Car" and "Norwegian Wood," the goofing had found a more disciplined musical focus. "Drive My Car" started life as a song about a "bitch" but evolved into a lyrical portrait of a wannabe "star of the screen," with its Lennonesque use of ambiguous story scraps which are steeped in sexual innuendo.

The girl of the song is trying to lure the singer ("I can show you a better time"), but, on his taking the job, she confesses to not actually having a car, but still wants him as her driver (letting him down gently with the very casual admission, "and *maybe* I'll love you"!) The track ends with the infectious backing vocals, "beep, beep, beep, beep, yeah," about which McCartney commented in *The Lyrics*, "It was always good to get nonsense lyrics in, and this song lent itself to

Rubber Soul (1965)

'beep, beep, beep, beep, yeah.' We did it in close harmony so it would sound a bit like a horn."

"Norwegian Wood"	Recorded: October 12/21, 1965	McCartney *(harmony vocals, bass guitar)*	Lennon *(double-tracked vocals, acoustic rhythm guitar)*
(Lennon-McCartney)	UK Release: December 3, 1965	Harrison *(double-tracked sitar)*	Starr *(tambourine, maracas, finger cymbal)*

On October 13, 1965, during the recording of "Norwegian Wood," McCartney jokingly told the *New Musical Express* that the band had found a new direction, "We've written some funny songs—songs with jokes in. We think that comedy numbers are the next thing after protest songs."

According to Lennon, "Norwegian Wood is . . . about an affair I was having. I was very careful and paranoid because I didn't want my wife, Cyn, to know that there really was something going on outside of the household. I'd always had *some* kind of affairs going, so I was trying to be sophisticated in writing about an affair, but in such a smoke-screen way that you couldn't tell."

"Norwegian Wood" marks milestones for the band. A roughly fifty-fifty Lennon and McCartney composition, it's the first Beatles song in which the lyric is more important than the music. Lennon's comment that "[he couldn't] remember any specific woman [the track] had to do with," the song's unusual instrumental texture, and its necessary lyrical ambiguity and obliqueness with regard to arson meant that its studied elusiveness found its way into a book of modern verse.

Another milestone is the track's use of an Indian instrument. Harrison had been interested in the sitar since his ears had pricked up when he'd heard the instrument provide exotic color to the soundtrack of *Help!* Lennon, too, at the time, became fascinated by the exotic raga phrases, which the song's descending melody may have been influenced by. The song is seen as a key work in the early evolution of world music.

McCartney says that the phrase "Norwegian Wood" is an ironic reference to the cheap pine wall paneling then in vogue in London. He also gave an amusing comment on the final lines of the song, "So I lit a fire, isn't it good, Norwegian wood?"

"I had this idea to set the Norwegian wood on fire as revenge," McCartney explained. "So we . . . burned the fucking place down as an act of revenge, and then we left it there and went into the instrumental."

"You Won't See Me"	Recorded: November 11, 1965	McCartney *(vocals, bass guitar, piano)*	Lennon *(backing vocals)*
(Lennon-McCartney)	UK Release: December 3, 1965	Harrison *(backing vocals, lead guitar)*	Starr *(drums, tambourine, hi-hat)*

Running to a length of three minutes and twenty-three seconds, McCartney's "You Won't See Me" was the band's longest track to date. It was inspired by the Tamla Motown sound. In his book, *Paul McCartney: Many Years From Now*, Barry Miles quotes McCartney as saying, "To me it was very Motown-flavored. It's got a James Jamerson feel. He was the Motown bass player, he was fabulous, the guy who did all those great melodic bass lines. It was him, me, and Brian Wilson who were doing melodic bass lines at that time, all from completely different angles, LA, Detroit and London, all picking up on what each other did." Indeed, *Rubber Soul* is when McCartney's melodic bass playing really begins to shine.

"Nowhere Man"	Recorded: October 21/22, 1965	McCartney *(harmony vocals, bass guitar)*	Lennon *(double-tracked vocal, acoustic rhythm guitar)*
(Lennon-McCartney)	UK Release: December 3, 1965	Harrison *(harmony vocals, lead guitar)*	Starr *(drums)*

"Nowhere Man" is symptomatic of the whole of *Rubber Soul*, which blends folk, rock, and soul, and explores more mature themes, with socially conscious lyrics.

Rubber Soul (1965)

Alan Pollack describes "Nowhere Man" as "a pioneering landmark example of what, within less than a year or so of its release, would be labeled as the 'folk rock' sub-genre," which "in spite of the electric arrangement and pop-ish choice of chords, an ingenuously simple tune and non-syncopated beat, help create a subtle fusion of styles."

Written by Lennon toward the latter end of recording *Rubber Soul*, when the band were pulling out all the stops to come up with enough tracks for the album, "Nowhere Man" was another product of his sense of isolation in his home in Weybridge, where many hours were spent in solitary contemplation away from the madness of now-waning Beatlemania.

"I was just sitting, trying to think of a song," Lennon told journalist Hunter Davies. "And I thought of myself sitting there, doing nothing and going nowhere," and, as quoted by David Sheff in *All We Are Saying*, "[Lennon] spent five hours that morning trying to write a song that was meaningful and good."

"I finally gave up and lay down," Lennon said. "Then 'Nowhere Man' came, words and music, the *whole* damn thing, as I lay down!"

Indeed, in March of the following year, Maureen Cleave wrote an article for London's *Evening Standard* in which she said, "[Lennon] can sleep almost indefinitely, is probably the laziest person in England. 'Physically lazy,' he said. 'I don't mind writing or reading or watching or speaking, but sex is the only physical thing I can be bothered with any more.'"

"Think For Yourself"	Recorded: November 8, 1965	McCartney *(harmony vocals, bass guitar)*	Lennon *(harmony vocals)*
(Harrison)	UK Release: December 3, 1965	Harrison *(vocals, lead guitar)*	Starr *(drums, maracas, tambourine)*

In 1978, George Harrison, along with American attorney and business partner Denis O'Brien, set up the British film production and distribution company HandMade Films to help finance the controversial Monty Python movie *Life of Brian*. During a crucial scene in the film, the religious faithful gather beneath a window *en masse* to receive

God's blessing, through Brian. This is the point at which Brian delivers the main message of the movie, "You don't need to follow anybody; you've got to think for yourselves!"

Thus it should come as no surprise that George was voicing the same message a dozen years earlier on *Rubber Soul*. In "Think For Yourself," George sings "Although your mind's opaque / Try thinking more if just for your own sake!" In his 1980 autobiography, *I Me Mine*, George wrote that the song "must [have been] written about somebody from the sound of it—but all this time later I don't quite recall who inspired that tune. Probably the government!"

"The Word"	Recorded: November 10, 1965	McCartney *(vocals, bass guitar, piano)*	Lennon *(vocals, rhythm guitar)*
(Lennon-McCartney)	UK Release: December 3, 1965	Harrison *(vocals, lead guitars)*	Starr *(drums, maracas)*

We find The Beatles here one step ahead of the still-undeveloped hippie counterculture. "The Word" finds the band singing for the first time about love as a concept. The song acts as a bridge between their idea of love in the likes of "Love Me Do" and its evolution into the universal notion of love expressed during, for example, their psychedelic-era track "All You Need Is Love."

"It sort of dawned on me that love was the answer, when I was younger, on the *Rubber Soul* album," Lennon was quoted in *Anthology*. "My first expression of it was a song called 'The Word.' The word is 'love,' in the good and the bad books that I have read, whatever, wherever, the word is 'love.' It seems like the underlying theme to the universe."

Indeed, in the song, Lennon admits that "in the beginning I misunderstood," but now he's "here to show everybody the light" and, if only we listeners do this wonderfully prosaic thing ("spread the word"), then miraculously "you'll be free" and, best-case scenario, "be like me."

In this way, the song foreshadows Lennon's later compositions on universal love. Moreover, one and a half years before the Summer of

Rubber Soul (1965)

Love and before the word "hippie" was even coined, to memorialize the arrival of this song, Lennon and McCartney shared a joint and created using crayons an artwork, which was "a psychedelic illuminated manuscript" of the "The Word's" lyrics.

Track recording details to look out for: the wacky piano intro, influenced by comedic piano pieces in the band-adored *Goon Show*, presages their imminent experimental period, and the definitively psychedelic bar-room piano which adorns the fading seconds of "Tomorrow Never Knows." McCartney's melodic and spontaneous bass-playing is inspired throughout.

"Michelle"	Recorded: November 3, 1965	McCartney *(vocals, bass guitar, acoustic guitars, lead guitar, drums)*	Lennon *(backing vocals)*
(Lennon-McCartney)	UK Release: December 3, 1965	Harrison *(backing vocals)*	Starr *(drums)*

Recorded in just nine hours, with playing and overdubs done by McCartney, this "equal part art song and neo-schmaltzy fox-trot" of a song would qualify as his first solo offering were it not for the backing vocals. McCartney explains in *The Lyrics* that the band was looking for new songs and that Lennon had asked McCartney, "Remember that daft French thing you used to do at parties?" Lennon was here referring to a lyric-less six-bar sliver of a melody which McCartney had sometimes busked in mock French to flirt with young women while hanging out with Lennon in Liverpool in the late 1950s.

"John was at art school by this time, so they had great art parties, and there'd be great girls," McCartney explains in *McCartney 3, 2, 1* to Rick Rubin. "So I would wear like a black turtle-neck sweater, and we were all trying to be French . . . I would sit in a corner and think, well if I wear the black turtle-neck and play guitar, then they might think I'm French!" McCartney explained that later he had met up with Ivan Vaughan, his best friend in school, who lived with his wife Jan in Islington.

"Jan taught French," McCartney said in *The Lyrics*, "so I asked if she could think of a rhyme for Michelle, two syllables. She said 'ma belle.' So, how could I say 'these words go together' in French? So, Jan also gave me 'sont les mots qui vont très bien ensemble.' You must sound the 'b' in 'ensemble.' I'd always said ensemble with a silent 'b' . . . I . . . grunted along like a cod Frenchman, and there was 'Michelle.'"

"What Goes On"	Recorded: 4 November 4, 1965	McCartney *(harmony vocals, bass guitar)*	Lennon *(harmony vocals, rhythm guitar)*
(Lennon-McCartney-Starkey)	UK Release: December 3, 1965	Harrison *(lead guitar)*	Starr *(vocals, drums)*

A near-parodic country-and-western ditty, sung by Ringo. Incidentally, why The Beatles' interest in country-and-western? One explanation may be that Irish music has had a significant influence on the development of country music, especially in the ballad style.

"I heard country-and-western music in Liverpool before I heard rock-and-roll." Lennon explained in a 1971 interview with *Rolling Stone*. Liverpool was known as a Nashville of the North in the 1950s, and as Lennon says, "The people there—the Irish in Ireland are the same—they take their country-and-western music very seriously. There's a big heavy following of it. There were established folk, blues, and country and western clubs in Liverpool before rock-and-roll and we were like the new kids coming out."

"Girl"	Recorded: November 11, 1965	McCartney *(backing vocals, bass guitar)*	Lennon *(vocals, acoustic guitars)*
(Lennon-McCartney)	UK Release: December 3, 1965	Harrison *(lead acoustic guitar)*	Starr *(drums)*

"Girl" was the final track laid down during the *Rubber Soul* sessions. The song partly explores the notion of the ideal woman, as Lennon explained in *The Beatles: In Their Own Words:* "This was about a dream girl. When Paul and I wrote lyrics in the old days, we used to laugh about it like the Tin Pan Alley people would. And it was only later on that we tried to match the lyrics to the tune. I like this one. It was one of my best."

With its decadent German two-step, its acoustic textures, and close vocal harmonies, "Girl" is a mix of musical influences.

Another *Rubber Soul* song qualifying for the "comedy" tag, and with the band's liking for innuendo, the middle section of "Girl" features Lennon and McCartney on backing vocals repeatedly singing the word "tit." As McCartney explained to Barry Miles, "The Beach Boys had a song out where they'd done 'la la la la' and we loved the innocence of that, and wanted to copy it, but not use the same phrase. So we were looking around for another phrase, so it was 'dit dit dit dit,' which we decided to change in our waggishness to 'tit tit tit tit' . . . George Martin might say, 'Was that "dit dit" or "tit tit" you were singing? Oh, "dit dit," George, but it does sound a bit like that, doesn't it?' Then we'd get in the car and break down laughing."

"I'm Looking Through You"	Recorded: October 24, November 6/11, 1965	McCartney *(double-tracked vocals, bass guitar, acoustic rhythm guitar, lead guitar)*	Lennon *(harmony vocals, acoustic rhythm guitar)*
(Lennon-McCartney)	UK Release: December 3, 1965	Harrison *(lead guitar?)*	Starr *(drums, tambourine, organ)*

Though a relatively minor track compared to the rest of the band's peerless back catalog, "I'm Looking Through You" is interesting for a number of reasons. Yet another rant about McCartney's tempestuous affair with Jane Asher, the song's instrumental texture is a fine example of the band's unique folk-rock style, dominated by the sound of acoustic guitars and electric bass.

The track, which probably started off as another nod to Bob Dylan, has bass, acoustic, and lead guitars all played by McCartney himself, while Ringo stabbed at a Vox Continental organ, and Lennon sang harmony vocal. It's perfectly possible Harrison doesn't appear on this track at all as, according to engineer Norman Smith, McCartney had become so perfectionist at this point that he had begun to insist on Harrison playing predetermined solos.

As recently as May 2024, Ringo said in an interview with *AXS TV* that, were it not for McCartney's work ethic, The Beatles would have made far fewer records. "Because of Paul, who was the workaholic of our band, we made a lot more records than John and I would've made.... We liked to sit around, Paul would call: 'Alright lads,' and we'd go in." During the *Rubber Soul* sessions, if Harrison failed to get the solos spot on, McCartney would simply play them himself and overdub, using a spare channel on the luxurious four-track recording deck. For the time being, Harrison kept his anger in check and his powder dry.

"In My Life"	Recorded: October 18/22, 1965	McCartney *(harmony vocals, bass guitar)*	Lennon *(double-tracked vocals, rhythm guitar)*
(Lennon-McCartney)	UK Release: December 3, 1965	Harrison *(harmony vocals, lead guitar)*	Starr *(drums, tambourine)*

Additional contributors: George Martin (piano)

Lennon's "In My Life" is a fascinating work of art. When Ozzy Osbourne spoke about the song in *Mojo* magazine in 2006, he said, "Lennon paints an incredible picture in just three minutes.... It's like when sometimes you sit down and think back to when you were a five-year-old and you were frightened when you went to school. These things come flooding back."

This nostalgic disquiet is partly explained by Alan Pollack: "The song creates a delicate and delicious balance between heart-baring intimacy of the first order and a vaguely subordinate and reticent unease. The closest I can pinpoint the latter is to something not quite straightforward about some of the chord progressions and the way in which the tune runs roughshod over them."

"In My Life" developed as Lennon's attempt to write a song about a bus ride from his childhood home at 250 Menlove Avenue into Liverpool's city center. As Lennon explained in 1980 to *Playboy*, "It was sparked by a remark a journalist and writer in England made after *In His Own Write* came out. . . . He said to me, 'Why don't you put some of the way you write in the book, as it were, in the songs? Or why don't you put something about your childhood into the songs?' . . . it was, I think, my first real major piece of work. Up till then it had all been sort of glib and throw-away. And that was the first time I consciously put my literary part of myself into the lyric."

"In My Life" is also typical of the way in which Lennon and McCartney's songwriting had evolved. Previously, before the influence of Bob Dylan, the band's songs might start with the lyrics, half-formed, then develop musically, and finally the remaining words would be ad-libbed along with the rest of the song in the studio; a kind of lyrics-melody-lyrics recipe. But by the time *Rubber Soul* came around, for Lennon in particular, the lyrics almost always preceded the music.

Rolling Stone ranked "In My Life" number 23 on its 2004 list of "The 500 Greatest Songs of All Time," number 98 on its 2021 list, and fifth on its list of the bands' "100 Greatest Songs." Over the last sixty years or so, fans and critics have pored over "In My Life," trying to define what makes the song so special. Perhaps it's the cryptic nature of Lennon's lyric, as they search for hidden meanings of his, and their own, life.

A footnote must be inserted here for the genius of the fifth Beatle, George Martin. In one of the musical highlights of the entire album, it was Martin who played and dubbed on that defining harpsichord-like electric piano part, recorded at half-speed to give it that inspired and nostalgic baroque sound.

"Wait"	Recorded: June 17, November 11, 1965	McCartney *(double-tracked vocals, bass guitar)*	Lennon *(double-tracked vocals, rhythm guitar)*
(Lennon-McCartney)	UK Release: December 3, 1965	Harrison *(guitars)*	Starr *(drums, tambourine, maracas)*

"Wait" is a fifty-fifty Lennon and McCartney composition that was written while the band were filming *Help!* in the Bahamas, and was originally intended for inclusion on that album, but was held back until the final sessions of the recording of *Rubber Soul*, when the imminent pre-Christmas deadline threatened to catch the band with a shortfall of new material.

The track has an appealing verve and demeanor, with a drive that stems from Ringo's drums, which shifts the rhythm and mood from the caustic tone of Lennon's verses to the more optimistic tone of McCartney's chorus and middle eight.

"If I Needed Someone"	Recorded: October 16/18, 1965	McCartney *(harmony vocals, bass guitar)*	Lennon *(harmony vocals, rhythm guitar)*
(Harrison)	UK Release: December 3, 1965	Harrison *(double-tracked vocals, lead guitar)*	Starr *(drums, tambourine)*

Additional contributors: George Martin (harmonium)

As the music of the 1960s progressed, a symbiotic relationship developed between some of the forerunner bands when it came to questions of sonic innovation. So, for example, George Harrison's twelve-string guitar work on "A Hard Day's Night" had played a crucial role in the evolution of the Rickenbacker-led jangly sound of American folk-rock band, The Byrds. And, in turn, Harrison credited The Byrds recordings "She Don't Care About Time" and "The Bells Of Rhymney" as being influential on "If I Needed Someone."

This track, the only one of George's songs that became part of the band's live repertoire in 1966, including their final show at San Francisco's Candlestick Park on August 29, is also a significant way-marker in revealing Harrison's growing interest in Indian music. The verses of "If I Needed Someone" contain an early example of a single-chord "sustained pedal" harmony, which would soon become George's MO for more unabashedly Indian tracks, such as "Blue Jay Way."

"Run For Your Life"	Recorded: October 12, 1965	McCartney *(harmony vocals, bass guitar)*	Lennon *(vocals, acoustic rhythm guitar)*
(Lennon-McCartney)	UK Release: December 3, 1965	Harrison *(harmony vocals, lead guitar)*	Starr *(drums, tambourine)*

This somewhat catchy track, which ends *Rubber Soul*, was actually recorded first. Based around a line from an Elvis Presley song, "Baby, Let's Play House," recorded in 1955, Lennon converts the original prosaic declaration of desire into an unfortunately menacingly sexist lyric, full of threatening, grabby jealousy. Lennon designated the track his "least favorite Beatles song" in a 1973 interview with *Rolling Stone*. The band never performed the song live.

REVOLVER (1966)
Psychedelia!

"The group encouraged us to break the rules. . . . It was implanted when we started Revolver *that every instrument should sound unlike itself: a piano shouldn't sound like a piano, a guitar shouldn't sound like a guitar. There were lots of things I wanted to try, we were listening to American records and they sounded so different, the engineers [at Abbey Road] had been using the same [methods] for years and years."*

—Geoff Emerick, *The Beatles: 10 Years That Shook the World, Mojo Magazine* (2004)

Revolver	Released: August 5, 1966	Recorded: April 6 – June 21, 1966	Duration: 35:01
Producer: George Martin	Studio: EMI, London	Label: Parlophone	Tracks: 14

Track Listing

Side One

No.	Title	Lead Vocals	Length
1	"Taxman"	Harrison	2:36
2	"Eleanor Rigby"	McCartney	2:11
3	"I'm Only Sleeping"	Lennon	3:02
4	"Love You To"	Harrison	3:00
5	"Here, There And Everywhere"	McCartney	2:29
6	"Yellow Submarine"	Starr	2:40
7	"She Said She Said"	Lennon	2:39

Side Two

8	"Good Day Sunshine"	McCartney	2:08
9	"And Your Bird Can Sing"	Lennon	2:02

(Continued)

Revolver (1966)

10	"For No One"	McCartney	2:03
11	"Doctor Robert"	Lennon	2:14
12	"I Want To Tell You"	Harrison	2:30
13	"Got To Get You Into My Life"	McCartney	2:31
14	"Tomorrow Never Knows"	Lennon	3:00

All songs written by Lennon-McCartney except tracks 4 and 12 written by Harrison

Psychedelia!

Once the tracks of *Rubber Soul* were finally in the can, The Beatles embarked on their final tour of Britain, ending with two gigs at the Capitol Cinema in Cardiff, Wales. The final itinerary was thankfully more fleeting when compared with earlier tours, and was followed by a ninety day break. It was the longest lay-off the band had had since 1962. John and Ringo jetted off "on vacation to the Trinidad Islands," George got married to Pattie Boyd, and Paul met Stevie Wonder at Stevie's performance at the Scotch of St. James nightclub in London. In short, to stay fresh and relevant and to be aware of any contemporary competition, the lads soaked up concepts and contemporary culture for their next album, *Revolver*.

The album's spawning during those days of 1966 chimed with the global recognition of London's role as a cultural capital. According to English author Philip Norman in his book *Shout! The Beatles in Their Generation*, the album *Revolver* captured the confidence of the summer of '66.

"It was hot pavements, open windows, King's Road bistros and England soccer stripes. It was the British accent, once again all-conquering."

The Swinging Sixties, that youth-driven cultural revolution focusing on modernity and hedonism, were in full force, with London at its center. Mod and psychedelic subcultures, the activism of anti-nuclear and feminist politics, and The Beatles were the multimedia leaders of the movement.

At the time, *Revolver* caught the attention of none other than American conductor and composer, Leonard Bernstein. In *Inside Pop: The Rock Revolution*, the 1967 documentary Bernstein made with David Oppenheim, Bernstein had special praise for *Revolver*: "These new adventures are simply extraordinary," he said, and he identified The Beatles' "eclecticism" as one of his favorite things about the album.

In Bernstein's words, the band felt the "freedom to absorb any and all musical styles and elements," citing the use of an elegiac string quartet in "Eleanor Rigby" as noteworthy and singled out George Harrison's "Love You To" as an homage to "the international and interracial way" The Beatles' music "ranges over the world, borrowing from the ragas of Hindu music."

Years later, in a Harvard lecture, Bernstein would go on to name "Eleanor Rigby" along with "A Day In The Life" and "She's Leaving Home" from *Sgt. Pepper's Lonely Hearts Club Band*, as pop songs worthy of being counted among the "great works" of the twentieth century.

Track by Track: *Revolver*

"Taxman"	Recorded: April 20-22, May 16, 1966	McCartney *(vocals, bass guitar, piano)*	Lennon *(backing vocals)*
(Harrison)	UK Release: August 5, 1966	Harrison *(vocals, lead guitar)*	Starr *(drums, cowbell, tambourine)*

"One, two, three, four; one two," against a bird-song backdrop of reverse tape loops. A smoker's cough. Then, suddenly, a gloriously thumping bassline and bass drum, closely followed by syncopated fuzz-toned guitar stabs, kicks the track into ebullient life. So begins "Taxman," the opening track of *Revolver*, one of the most crucial albums in rock history.

The band had always gone for a statement song as an album opener; something loud, hard-hitting, and impactful. Think "I Saw Her Standing There" on *Please Please Me*, "It Won't Be Long" on *With*

Revolver (1966)

The Beatles, and "No Reply" on *Beatles For Sale*. *Revolver* presents George Harrison's one-time-only stab at that opening track position.

Penned with a little lyrical help from Lennon, Harrison's "Taxman" protested the level of progressive tax imposed by the contemporary government, which, to George's horror, saw the band paying a 95 percent supertax; hence the lyric "There's one for you, nineteen for me." Harrison had recently learned that The Beatles' tax burden might eventually lead to bankruptcy for some artists, and he was militant in his opposition to the government using the band's income to help fund the production of military weapons (governments never change, it seems). The track depicts the untiring taxman pursuing revenue irrespective of political color, as we hear Lennon's backing vocal sing about "Labour" leader Harold Wilson as well as Conservative leader, Edward Heath. Not only was "Taxman" the band's first topical song but it was also their first overt political statement in music.

The texture of the track comprises McCartney's remarkably banging bass against the off-beat spike of fuzzed-up guitar, and Ringo's cowbell which counts the "coins" as they fall into the coffers of the Treasury. Check out McCartney's startling guitar solo, so good that Alan Pollack described it as sounding like "Clapton's own handiwork, only sped up to the frantically comical pace of the Keystone Cops," referring to the frenetic sound of McCartney's guitar phrasing.

"Eleanor Rigby"	Recorded: April 28-29, June 6, 1966	McCartney *(vocals)*	Lennon *(harmony vocals)*
(Lennon-McCartney)	UK Release: August 5, 1966	Harrison *(harmony vocals)*	

Additional contributors: Unnamed musicians played violins, violas, and cellos

John Lennon had been the band's dominant creative force before *Revolver*. Paul now attained parity. And what a start was "Eleanor Rigby." At the time *Revolver* was recorded, pop music had rarely dealt with loneliness and death. Now, compositions which departed from the mainstream drew in droves of new and serious-minded listeners.

According to McCartney in *The Lyrics*:

> The song itself was consciously written to evoke the subject of loneliness, with the hope that we could get listeners to empathize. Those opening lines, "Eleanor Rigby/Picks up the rice in the church where a wedding has been/Lives in a dream" ... I wanted to make it more poignant than her just cleaning up afterwards, so it became more about someone who was lonely. Someone not likely to have her own wedding, but only the dream of one. Allen Ginsberg told me it was a great poem, so I'm going to go with Allen. He was no slouch.

After some considerable evolution of the lyrics (for example, "Father McCartney" to "Father McKenzie" in case the vicar was mistaken for Paul's dad) and having only the melody and first verse down, McCartney presented the song to the rest of the band in the music room of Lennon's home. Once Harrison had come up with the "all the lonely people" hook and Ringo had offered the "writing the words of a sermon that no one will hear" line, McCartney had a string component in mind for the song's recording, especially after the huge success of the string quartet on "Yesterday."

On taking his ideas to George Martin, McCartney explained that he wanted a series of string chord stabs as backing to the vocals.

"In George's version of things," McCartney explains in *The Lyrics*, "he conflates my idea of the stabs and his own inspiration by Bernard Herrmann, who had written the music for the movie *Psycho*. George wanted to bring some of that drama into the arrangement. And, of course, there's some kind of madcap connection between Eleanor Rigby, an elderly woman left high and dry, and the mummified mother in *Psycho*."

Revolver (1966)

"I'm Only Sleeping"	Recorded: April 27/29 and May 5/6, 1966	McCartney *(harmony vocals, bass guitar)*	Lennon *(double-tracked vocals, acoustic rhythm guitar)*
(Lennon-McCartney)	UK Release: August 5, 1966	Harrison *(harmony vocals, lead guitar)*	Starr *(drums)*

As we have seen, Lennon spent the majority of 1966 experimenting daily with the psychedelic drug, LSD. Deprived of the routine that came with the band's rigorous touring schedule, he was content to pass away the time tripping at home.

Lennon's "I'm Only Sleeping" was inspired by McCartney's "annoying" habit of waking him up from his slumber for their afternoon songwriting sessions at Lennon's abode. The band worked hard to create the sonic, dreamlike quality of the song by recording the basic track at a faster tempo and then slowing it down using varispeed as they'd done on the track "Rain" to give a thick and "heavy" sound to the production. Multitracking was applied to Lennon's voice to replicate, as one critic put it, a "papery old man's voice."

Another fascinating feature of the song is the backwards guitar solo played by Harrison. It's actually comprised of two separate parts, created and performed by Harrison during a late-night session that lasted six hours in May 1966. Harrison wrote down the Indian-flavored notation that he had created for the solo. George Martin transcribed it in reverse, and recorded it like that twice: once with fuzz effects and the other "clean." The track was then subsequently dubbed on with the tape running backwards to create the sonic landscape.

According to McCartney in *Many Years From Now,* the idea for the backwards solo came after Lennon had loaded the tape into the tape machine incorrectly.

"It played backwards, and, 'What the hell is going on?' Those effects! Nobody knew how those sounded then. We said, 'My God, that is fantastic! Can we do that for real?' . . . So that was what we did and that was where we discovered backwards guitar. It was a beautiful solo actually. It sounds like something you couldn't play."

"Love You To"	Recorded: April 11/13, 1966	McCartney *(harmony vocals)*	*(Various unnamed Indian musicians)*
(Harrison)	UK Release: August 5, 1966	Harrison *(multi-tracked vocals, acoustic guitar, electric guitar)*	Starr *(tambourine)*

In the wake of Lennon's "Norwegian Wood" on *Rubber Soul*, George's "Love You To" was the band's next attempt to better understand and perform Indian music, in this case in the classical Indian style. As Harrison explained, "I wrote 'Love You To' on the sitar, because the sitar sounded so nice and my interest was getting deeper all the time. I wanted to write a tune that was specifically for the sitar. Also, it had a tabla part, and that was the first time we used a tabla player."

And so, under the working title of "Granny Smith," George's "Love You To" was born. On two days in April of 1966, the band built layers of takes, one with George singing with an acoustic, another with Paul on harmony vocal, and others which featured the sitar, tabla, bass, and fuzz guitars.

Thus, it was that The Beatles contributed to this sudden flash of interest in Indian music at the time, along with Brian Jones of The Rolling Stones using a sitar on "Paint It Black," and Dave Mason of Traffic using a sitar on the band's 1967 hits "Paper Sun" and "Hole In My Shoe." Sadly, while Indian music is pleasingly seductive and psychedelic, it is not an easily acquired taste for Western ears, built as it is on unfamiliar techniques and reflective of a different philosophical outlook.

"Here, There And Everywhere"	Recorded: June 14/16/17, 1966	McCartney *(multi-tracked vocals, acoustic guitar, bass guitar, finger snaps)*	Lennon *(backing vocals, finger snaps)*
(Lennon-McCartney)	UK Release: August 5, 1966	Harrison *(backing vocals, lead guitar, finger snaps)*	Starr *(drums, finger snaps)*

Even more than "Yesterday," "Here, There And Everywhere" is a personal favorite of McCartney's. The song boasts a brief prelude: "To lead a better life, I need my love to be here," which McCartney confesses is an homage to the kind "rambling preamble," as he puts it, of some of Cole Porter's songs.

The song has a sublime circularity. As McCartney says in *The Lyrics*, "What I like most about this song is that we think we're on a path on the moors and we're going for a walk, and then suddenly we've arrived where we started. It's not quite that we've gone around in a circle. It's more magical than that. We've come to another beginning of the path. You can see back to where you came from, and you're definitely not there. You're in a new place, though it's got the same scenery. I've always liked that trick."

In terms of circularity, McCartney was pleased by the fact that "Here, There And Everywhere" was influenced by the very recent recording of "God Only Knows" on the Beach Boys album *Pet Sounds*, and that "God Only Knows" itself was inspired by Brian Wilson's listening to The Beatles songs on *Rubber Soul*.

"Yellow Submarine"	Recorded: May 25, June 1, 1966	McCartney *(backing vocals, bass guitar, shouting)*	Lennon *(backing vocals, acoustic guitar, shouting)*
(Lennon-McCartney)	UK Release: August 5, 1966	Harrison *(backing vocals, tambourine)*	Starr *(vocals, drums)*

According to former Nirvana drummer and Foo Fighters leader Dave Grohl, "Yellow Submarine" is the "song [that] has defined everyone in the world's life at some point." Dave's idea is that each of us has an inbuilt playlist of songs that have marked out our lives in some way. Perhaps a tune to which you rocked around the living room as a kid, or maybe a melody that accompanied your first kiss—songs that acted as a soundtrack to your life.

"Yellow Submarine" is rare in that it has the charm to urge anyone, young or old, to sing along to its lyrics, according to Grohl. Beautifully prosaic, jolly, and melodic, the song appeals to all demographics. (One of your authors [Mark Brake] has a fond and sunny memory of driving across a seven-mile stretch of sand on the south coast of Wales, with kids and cousins in the back of the charabanc, as we all sang along to Ringo one distant summer in the late nineties. One struggles to imagine a similar scene in the future with songs such as "Smack My Bitch Up.")

"Yellow Submarine" is quintessential Beatles. Embedding a newfound psychedelic vibe into a song about sailing up to the sun in a submersible, the band magically transforms an otherwise throwaway novelty song into a vehicle for psychedelia itself. Like a psychotropic drug, "Yellow Submarine" affects the listener's mind and mood.

According to McCartney in *The Lyrics*, the song's inspiration came from a number of places. Chief among these was firstly his school experience of the "nonsense tradition" of English prose, such as Lewis Carroll, and poetry, such as Coleridge's "The Rime of the Ancient Mariner."

"A large part of the subtext of "Yellow Submarine" was that [nonsense tradition], even then, The Beatles were living in our own capsule. Our own microclimate. Our own controlled environment," McCartney says.

Another source of inspiration was science exploration programs on TV.

"That can't be overstated," McCartney says again in *The Lyrics*. "The incredible popularity of television programs at the time that featured the underwater world. There were the Austrian underwater divers Hans and Lotte Hass. Lotte was a kind of pin-up. . . . That underwater world was quite magical."

Such science programs also opened up new possibilities for a generation that grew up with post-war experiences.

"When we were kids in Liverpool—'*In the town where I was born*'— it was all bombs and rationing and ruins," McCartney says. "So, when you got a bit more, it was like going from black and white to color. For The Beatles—though we didn't know it at the time—expressing our joy at coming out of the black-and-white world actually contributed to that new burst of color. It's hard to believe, but we played an active role in it. We helped to make the '*Sky of blue and sea of green*' so vibrant."

Revolver (1966)

Taking twice as much studio time as the band's entire debut album, *Please Please Me*, the sheer ambition of "Yellow Submarine" was achieved by firstly laying down a simple track. (In October of 2022, as part of the launch of a new special edition of the band's *Revolver* album, a previously unreleased early demo was released which features a sparse and melancholic acoustic early version of the verse's melody from Lennon. As *Rolling Stone* said at the time, "The Beatles could pack an emotional punch like no other band . . . it's one of the biggest surprises: who expected emotional depth from 'Yellow Submarine'? But like so many moments on the new edition, 'Yellow Submarine' makes you rethink everything you thought you knew about the group. It shows how far they were willing to experiment on *Revolver*, pushing out of their comfort zones. . . . It all captures the freewheeling spirit of the *Revolver* sessions—four boys running wild in the clubhouse, inventing the future.")

The basic track then acted as a firm foundation for the goofy-but-futuristic collage of sampled soundbites that overlay it, which the lads themselves took great trouble to synthesize. They enlisted an army of guests to help them, including Pattie Boyd, Brian Jones of The Stones, along with Mick Jagger's then-partner, Marianne Faithfull, the band's road managers, Neil Aspinall and Mal Evans, and The Beatles' driver, Alf Bicknell.

The "Goonish concerto" of sound effects (sourced from the studio store cupboard and including bells, chains, hooters, whistles, a tin bath, and a cash till) rests heavily on George Martin's experience of producing British comedy records for *Beyond the Fringe* and *The Goons*. The sound of ocean waves, which can be heard at the start of the second verse and through the first chorus, was created by George Harrison swirling water around a bathtub. The party atmosphere in the second verse was made through a fusion of snatches of excited chatter, Patti Boyd's piercing shrieks, Brian Jones clinking glasses and blowing on an ocarina, and Alf Bicknell adopting an Ebenezer Scrooge persona in the rattling of chains, and the tumbling of coins.

To produce the brass cacophony following the line "and the band begins to play," George Martin, along with studio engineer Geoff Emerick repurposed a copy of a brass band recording from EMI's tape library, which they disguised by splicing up the tape and reshuffling the melody.

Finally, in the last verse, Lennon echoes Ringo's lead vocal, delivering lines in a manner that musicologist Walter Everett calls "manic." Sadly, Walter seems to be unaware of the band's British influences, as Lennon's delivery is typical of the kind of pure madcap tomfoolery of sound effects heard in *The Goon Show*. Very keen to sound as if he were singing underwater, Lennon tried taping his vocal part with a microphone sheathed in a condom and submerged inside a bottle filled with water. When this beautifully creative solution proved ineffective, Lennon simply sang with the microphone plugged through a Vox guitar amplifier. When the overdubs were done, the "Yellow Submarine" sessions ended up with everybody in a line doing a conga dance around the studio while Mal Evans banged on a big marching bass drum that was strapped to his chest. George Martin later told the *NME* that the band "loved every minute" of the session and that the experience was "more like the things I've done with *The Goons* and Peter Sellers" than a Beatles recording.

As Alan Pollack points out, "It's also worth recalling just what an attention-grabbing curve ball this song appeared to be in context of its initial release. Sure, The Beatles had been growing ever more difficult to pigeon-hole for a while by mid-1966, but the appearance of this song (three days after *Revolver*, separately released as the B-side of the single "Eleanor Rigby"), no less, promised to go the limit. Could anyone other than The Beatles get away with this? Try to imagine 'Yellow Submarine' as the first or second song of a no-name group." Peerless.

Revolver (1966)

"She Said She Said"	Recorded: June 21, 1966	McCartney ?	Lennon (*vocals, rhythm guitar, Hammond organ*)
(Lennon-McCartney)	UK Release: August 5, 1966	Harrison (*backing vocals, lead guitar*)	Starr (*drums, shaker*)

This song is about a meeting between The Beatles and The Byrds, In the summer of 1965, the two bands met up. The Beatles were on their US tour and had rented a house in Los Angeles' Mulholland Drive. So, on August 24, while members of the band played host to Roger McGuinn and David Crosby, and the quintet of musicians, Paul was absent and spent the day experimenting with LSD. In due course, the actor Peter Fonda arrived at the house, also tripping on acid, and promptly tried to reassure Harrison, who felt like he was dying.

"I told him there was nothing to be afraid of and that all he needed to do was relax. I said that I knew what it was like to be dead because when I was ten years old, I'd accidentally shot myself in the stomach and my heart stopped beating three times while I was on the operating table because I'd lost so much blood. John was passing at the time and heard me saying 'I know what it's like to be dead.' He looked at me and said, 'You're making me feel like I've never been born. Who put all that shit in your head?'"

Lennon explained his reaction in a 1980 interview, "We didn't *want* to hear about that! We were on an acid trip and the sun was shining and the girls were dancing and the whole thing was beautiful and Sixties, and this guy—who I really didn't know; he hadn't made *Easy Rider* or anything—kept coming over, wearing shades, saying, 'I know what it's like to be dead,' and we kept leaving him because he was so boring! . . . It was scary. You know . . . when you're flying high and [*whispers*] 'I know what it's like to be dead, man.'"

The last track laid down for the album, "She Said She Said" is one of the most rhythmically complex things Lennon ever wrote, and most of the nine-hour session used to get it in the can was spent on rehearsal. The final product justifies every minute. The band perform brilliantly, from Harrison's help on vocal harmonies and lead guitar, to Ringo's tour de force in fine technical drumming. The overall effect is an outstanding track as emotionally tense and as moving in its own melancholic way as "Eleanor Rigby." McCartney was quoted in *Many Years From Now* as saying, "John brought ["She Said She Said"] in pretty much finished, I think. I'm not sure but I think it was one of the only Beatle records I never played on. I think we had a barney or something and I said, 'Oh, fuck you!' and they said, 'Well, we'll do it.' I think George played bass [actually additional bass notes were played on a Hammond organ after McCartney departed.]"

"Good Day Sunshine"	Recorded: June 8/9, 1966	McCartney *(vocals, piano, claps)*	Lennon *(harmony vocals, claps)*
(Lennon-McCartney)	UK Release: August 5, 1966	Harrison *(harmony vocals, bass guitar, claps)*	Starr *(drums, claps)*

Perfectly encapsulating the hot summer of '66, "Good Day Sunshine" shows the band, and especially its author McCartney, at their effortless best—the kind of effortlessness that competitors could only dream of. McCartney wrote the song at Lennon's house in Weybridge during one of their afternoon songwriting sessions.

"It was really very much a nod to The Lovin' Spoonful's 'Daydream,'" according to McCartney in *Many Years From Now*. "The same traditional, almost trad-jazz feel. That was our favorite record of theirs. 'Good Day Sunshine' was me trying to write something similar to 'Daydream.' John and I wrote it together at Kenwood, but it was basically mine, and he helped me with it."

Recorded quickly in three takes over just two days, with minimal studio trickery (apart from the usual vari-speed piano played by Martin), the track features George on bass while McCartney takes on the main piano duties and delivers a sublime vocal performance. The track is noticeable for its lack of guitar parts, and the stunning cascade of voices fading out in the coda.

Cosmic science fact: The NASA "Chronology of Wakeup Calls" shows that McCartney played the song live to the astronauts aboard the International Space Station on November 13, 2005, in the first concert link-up to the space station. The Beatles remain stellar.

"And Your Bird Can Sing"	Recorded: April 20/26, 1966	McCartney *(harmony vocals, bass guitar, claps)*	Lennon *(vocals, rhythm guitar, claps)*
(Lennon-McCartney)	UK Release: August 5, 1966	Harrison *(harmony vocals, lead guitar, claps)*	Starr *(drums, tambourine, claps)*

Graced with one of their most intricate guitar riffs, played jointly by McCartney and Harrison, "And Your Bird Can Sing" was not fondly remembered by Lennon, its chief author. According to David Sheff, Lennon described it as "another of my throwaways... fancy paper around an empty box."

His description was as cryptic as the lyrics, perhaps a nod to the influence that Dylan was having on him at the time. Lennon never publicly discussed the song in detail, so a myriad of interpretations have sprung up around its meaning. The most popular interpretation is that the song is a jibe at Mick Jagger of The Rolling Stones, who Lennon saw as Beatles copyists. The "bird" in this context would be the Scouse slang word for "girlfriend." Musically the song was initially recorded with a Byrds-type sound defined by the jangly Rickenbacker 360 Deluxe electric twelve-string, which the Byrds themselves copied from The Beatles back in 1964.

An early version can be heard on *Anthology 2*, and contains giggling from Lennon and McCartney, spoken parts and whistling along with the melody. What fun these boys had in the studio. This version was revamped a few days later, with the jangly guitars being replaced with the heavier sound of the twin Epiphone Casinos. Genesis drummer Phil Collins described the song as "one of the best songs ever written, and it's only a minute and a half long."

"For No One"	Recorded: May 9/16/19, 1966	McCartney *(vocals, bass guitar, piano, clavichord)*	
(Lennon-McCartney)	UK Release: August 5, 1966		Starr *(drums, tambourine)*

Additional contributors: Alan Civil (horn)

"For No One," according to McCartney in *The Lyrics*, is "a song about rejection.... It's a horrible moment when you've broken up with someone, and you look at them—this person you used to be in love with, or *thought* you were in love with—and none of that old feeling is there."

Heartache in love is always a rich area to explore in song, and "For No One" is approached, according to Alan Pollack, "by borrowing elements of the early nineteenth century 'Art Song'; I place this one on the Spectrum of Style somewhere in between 'Eleanor Rigby' and 'Michelle.'"

The song started while McCartney was on a skiing holiday in Switzerland, and evolved once more along the lines of McCartney and George Martin seeking some new kind of instrumentation to enhance the song's feel. Very often, the band's approach was musically intuitive, rather than formal and methodological, so much of the science of their tradecraft was in the way George Martin used the available recording technology to realize that gut-felt intuition.

"I was interested in the French horn, because it was an instrument I'd always loved from when I was a kid," McCartney explained in *Anthology*. Martin wrote down the delicate melody that McCartney sang to him and, forever pushing boundaries, the pair decided to insert a top note into the score outside the horn's normal range.

"We came to the session and Alan [Civil, the French horn session musician] looked up from his bit of paper: 'Eh, George? I think there's a mistake here—you've got a high F written down.' Then George and I said, 'Yeah,' and smiled back at him, and he knew what we were up to and played it. These great players will do it."

Revolver (1966)

| "Doctor Robert" | Recorded: April 17/19, 1966 | McCartney *(harmony vocals, bass guitar)* | Lennon *(vocals, rhythm guitar, harmonium)* |
| (Lennon-McCartney) | UK Release: August 5, 1966 | Harrison *(lead guitar, maracas)* | Starr *(drums)* |

Written mostly by Lennon, "Doctor Robert" is notable for containing the band's first explicit reference to drugs, though at the time the allusions largely went unnoticed. For some, the song is about a shifty New York medic, a Dr. Robert Freymann, who ran a discreet clinic on Manhattan's East 78th Street. The doctor allegedly got his well-heeled socialite clients hooked on amphetamines by creatively mixing their B-12 vitamin shots. Thus, the song shifts key evasively, settling down only in the middle eight, where the good doctor, coming over as a clinical version of a snake-oil salesman, waxes evangelical about his wares, as somewhere a harmonium plays, and the other Beatles do their best to sound like choirboys.

Others suggest the song is Lennon's subconscious sideways swipe at the trustworthiness of "Doctor" Timothy Leary, while Lennon himself, probably covering his own tracks, claimed it was an autobiographical song. As Lennon is quoted as saying in David Sheff's *All We Are Saying*, "Mainly about drugs and pills. It was about myself. I was the one that carried all the pills on tour. Well, in the early days. Later on, the roadies did it. We just kept them in our pockets loose. In case of trouble."

However, according to McCartney, in *Many Years From Now*, "John and I thought it was a funny idea: the fantasy doctor who would fix you up by giving you drugs, ["Doctor Robert"] was a parody on that idea. It's just a piss-take.... Change your blood and have a vitamin shot and you'll feel better."

The caustic behavior of the doctor is matched by Lennon's acerbic vocal, which is cleverly married to McCartney's choirboy harmony ("he's a man you must believe"), and by Harrison's double-tracked guitar, which is a creative fusion of country-and-western and sitar.

"I Want To Tell You" (Harrison)	Recorded: June 2/3, 1966	McCartney *(harmony vocals, bass guitar, piano, claps)*	Lennon *(harmony vocals, tambourine, claps)*
	UK Release: August 5, 1966	Harrison *(double-tracked vocals, lead guitar, claps)*	Starr *(drums, maracas, claps)*

Revolver is the only Beatles album to feature three songs written by George. Harrison's writing had really begun to blossom by early 1966, no doubt due to the band having more creative space as the result of a lull in their professional commitments. Harrison's songs reveal a growing maturity as he starts to develop his own musical identity, bolstered by his increasing interest in Indian music and culture. "I Want To Tell You" reveals the dual philosophical influence of Indian music and LSD. In his autobiography, *I, Me, Mine*, George says that the song addresses "the avalanche of thoughts that are so hard to write down or say or transmit."

The stuttering guitar riff of this song and the dissonance in the melody perfectly reflect the erratic communication to which the song refers. The song starts with a fade-in, in the same vein as "Eight Days A Week," with McCartney on piano and Harrison's guitar piece being played through a Leslie speaker. The backing vocals from McCartney are a joy to behold, particularly his use of Indian-style vocal melisma (the singing of a single syllable of text while moving between several different notes in succession) in the fade out. This track also marks the first time that McCartney recorded his bass part after the band had completed the rhythm track, a habit that became commonplace in the band's subsequent work.

"Got To Get You Into My Life" (Lennon-McCartney)	Recorded: April 7/8/11, May 17/20, June 1966	McCartney *(double-tracked vocals, bass guitar)*	Lennon *(rhythm guitar)*
	UK Release: August 5, 1966	Harrison *(lead guitar)*	Starr *(drums, tambourine)*

McCartney's ode to marijuana, "Got To Get You Into My Life," is a Motown-influenced track, written after seeing Stevie Wonder perform in London in February 1966. The song showcases a strong vocal performance from McCartney and a big brass sound played by two members of Georgie Fame's group The Blue Flames, whom John and Paul knew from the London club scene.

To get the big sound for the brass section, the microphones were placed in the bells of the instruments. These horn parts were then duplicated with a slight delay, effectively double tracking the parts. This was the first time The Beatles had used a horn section on one of their songs—an interesting insight into how fast the band experimented in the studio can be heard on *Anthology 2*.

The song seems to have gone through a series of changes; the early version is slower with acoustic guitars, a harmonium and an a-cappella section. Harrison added guitar parts before the "masters of flow" decided on the glorious brass arrangement that made it onto *Revolver*.

"Tomorrow Never Knows"	Recorded: April 7/8/11, May 18, June 17/20, 1966	McCartney *(bass guitar, tape loops)*	Lennon *(vocals, organ, tape loops)*
(Lennon-McCartney)	UK Release: August 5, 1966	Harrison *(guitars, sitar, tambura, tape loops)*	Starr *(drums, tambourine, tape loops)*

Additional contributors: George Martin *(piano)*

In 1966, no one had ever heard anything like "Tomorrow Never Knows." Over fifty years later and it still sounds like nothing else. As Tom Rowlands of The Chemical Brothers explained in *Mojo* in 2006, "[We] always used to play it after an acid house track . . . and it'd be incredible. People would ask if it was something new, or a remix—it just sounded so intense and so wild. I've had stages of my life when I've been completely obsessed with that song. It's immediate, but at the same time it's incomprehensible. It's noise, but it's music. The idea that a screech could be the hook of a song was revolutionary."

"'Tomorrow Never Knows' is still to be bettered, in terms of forward-looking, futuristic sound," Paul Weller commented in *The Observer* in May 2024.

The song remains a monumental achievement for the band, paving the way for the extensive experimentation which came later, especially on songs such as "Strawberry Fields Forever" and "I Am The Walrus." "Tomorrow Never Knows" is one of their most innovative tracks. The song was the first to be recorded for *Revolver*, with Lennon taking lyrical inspiration from Timothy Leary's *The Psychedelic Experience*, the book itself an adaptation of the ancient *Tibetan Book of the Dead*. Leary's book was intended as a guidebook for those, such as Lennon and Harrison, seeking spiritual enlightenment through the use of psychedelic drugs.

The song title was another one of Ringo's sayings, as Lennon explained in David Scheff's *All We Are Saying*, "That's me in my *Tibetan Book of the Dead* period. I took one of Ringo's malapropisms as the title, to sort of take the edge off the heavy philosophical lyrics."

The song itself is based around a single chord, a direct result of their developing interest with Indian music which pervades the album. The recording of the track contains a remarkable amount of technical studio innovation. In the months prior to recording *Revolver*, all four Beatles had bought reel-to-reel tape recorders.

According to studio engineer Geoff Emerick in his *Here, There and Everywhere*, "Paul had discovered that the erase head could be removed, which allowed new sounds to be added to the existing ones each time the tape passed over the record head. Because of the primitive technology of the time, the tape quickly became saturated with sound and distorted."

These revolutionary tape loops would play an integral part in the recording. The song is driven along by Ringo's thunderous, hypnotic drum part. Emerick was responsible for the new drum sound by slackening the tom tom skins, stuffing an old woolen jumper inside the bass drum and moving the drum microphones up very close. The recording of the drum part was heavily compressed and echoed to give an extraordinary sound. McCartney's bass line closely follows the drums to enhance the hypnotic effect.

In total, there were six tape loops used on "Tomorrow Never Knows:" a "seagull" noise, which is actually a recording of McCartney laughing; an orchestra playing a B flat chord; notes from on a Mellotron's flute setting (a mellotron is an electro-mechanical musical instrument with various settings that mimic the sounds of acoustic instruments); a second Mellotron on its violin setting; a finger rubbing the rim of a wine glass; and a distorted sitar which is most clearly heard in the instrumental break following the lines, "It is being, it is being." For the recording of Lennon's vocals, manual double-tracking was used, and for the second part, the Abbey Road engineers came up with an innovative solution to give Lennon the sound he had heard in his head.

As George Martin explained in *Anthology*, "For 'Tomorrow Never Knows' he said to me he wanted his voice to sound like the Dalai Lama chanting from a hilltop. I knew perfectly well that an ordinary echo or reverb wouldn't work, because it would just put a very distant voice on. We needed to have something a bit weird and metallic."

To achieve the right sound, the vocal was fed through a rotating Leslie speaker, whose speed of rotation could be varied.

"By putting his voice through that and then recording it again, you got a kind of intermittent vibrato effect, which is what we hear on 'Tomorrow Never Knows.' I don't think anyone had done that before. It was quite a revolutionary track for *Revolver*."

In April 1966, the basic track was recorded in three takes. The band then laid various tape loops over the track. As George Martin explained, this was done live in the studio to produce the unique recording.

"We did a live mix of all the loops. All over the studios we had people spooling them onto machines with pencils while Geoff Emerick did the balancing."

Geoff Emerick continues the story, "George Martin and I huddled over the console, raising and lowering faders to shouted instructions from John, Paul, George, and Ringo. ('Let's have that seagull sound now!' 'More distorted wine glasses!') With each fader carrying a different loop, the mixing desk acted like a synthesizer, and we played it like a musical instrument, too, carefully overdubbing textures to the pre-recorded backing track."

The result is something almost impossible to copy.

"It is the one track, of all the songs The Beatles did, that could never be reproduced," George Martin explained. "It would be impossible to go back now and mix exactly the same thing: the 'happening' of the tape loops, inserted as we all swung off the levers on the faders willy-nilly, was a random event."

"Tomorrow Never Knows," like *Revolver* itself, is a remarkable achievement, especially when one considers it was all done through the experimentation of mere four-track recording technology, while showcasing such brilliant, innovative minds.

Sgt. Pepper's Lonely Hearts Club Band (1967)

Art Rock

We were fed up with being "The Beatles." We really hated that fucking four little mop-top approach. We were not boys, we were men ... and thought of ourselves as artists rather than just performers.
    —Paul McCartney, quoted in Barry Miles' *Paul McCartney: Many Years From Now* (1997)

Sgt. Pepper's Lonely Hearts Club Band	Released: June 1, 1967	Recorded: December 6, 1966 – April 21, 1967	Duration: 39:36
Producer: George Martin	Studio: EMI and Regent Sound, London	Label: Parlophone	Tracks: 13

Track Listing

Side One

No.	Title	Lead Vocals	Length
1	"Sgt. Pepper's Lonely Hearts Club Band"	McCartney	2:00
2	"With A Little Help From My Friends"	Starr	2:42
3	"Lucy In The Sky With Diamonds"	Lennon	3:28
4	"Getting Better"	McCartney	2:48
5	"Fixing A Hole"	McCartney	2:36
6	"She's Leaving Home"	McCartney with Lennon	3:25
7	"Being For The Benefit Of Mr. Kite!"	Lennon	2:37

Side Two

8	"Within You Without You"	Harrison	5:05
9	"When I'm Sixty-Four"	McCartney	2:37
10	"Lovely Rita"	McCartney	2:42

(Continued)

Sgt. Pepper's Lonely Hearts Club Band (1967)

11	"Good Morning Good Morning"	Lennon	2:42
12	"Sgt. Pepper's Lonely Hearts Club Band (Reprise)"	Lennon, McCartney, Harrison and Starr	1:18
13	"A Day In The Life"	Lennon with McCartney	5:38

All songs written by Lennon-McCartney except track 8 written by Harrison

Art Rock

On June 1, 1967, The Beatles released an album which redefined what a rock album could or should be. *Sgt. Pepper's Lonely Hearts Club Band* was immediately acclaimed as a revolution in popular music. The vast majority of contemporary reviews of the album were hugely positive. The influential *Village Voice* called *Sgt. Pepper* "the most ambitious and most successful record album ever issued." In Britain, the *Record Mirror* said *Sgt. Pepper* was "clever and brilliant, from raucous to poignant and back again," while *Disc and Music Echo* called it "a beautiful and potent record, unique, clever, and stunning." The London *Times* described *Sgt. Pepper* as a "pop music master-class" and suggested that, so marked were its musical advances, "the only track that would have been conceivable in pop songs five years ago" was "With A Little Help From My Friends."

Meanwhile, *Gramophone* magazine identified the quintessential elements of The Beatles that made *Sgt. Pepper* such a success, with "the combination of imagination, cheek and skill" that made the new album "like nearly everything the Beatles do, bizarre, wonderful, perverse, beautiful, exciting, provocative, exasperating, compassionate and mocking." Little wonder that, in *The Oxford Encyclopedia of British Literature*, American cultural critic Kevin Dettmar writes that *Sgt. Pepper* achieved "a combination of popular success and critical acclaim unequaled in twentieth-century art . . . never before had an aesthetic and technical masterpiece enjoyed such popularity."

Nothing like *Sgt. Pepper* had been before.

Sgt. Pepper was *the* album of 1967's Summer of Love, and its music was psychedelic and revolutionary. Maybe more than any Beatles album before, *Sgt. Pepper* chimed with the times; of the thirteen songs on the album, over half were inspired by newspaper stories, photographs, or quotidian events. The album had an instant cross-generational impact and was related to various touchstones of 1960s youth culture, including fashion, drugs, mysticism, and the decade's sense of optimism and empowerment.

The other main factor in the album's huge historical influence was its sheer musical ambition and scope. All musical styles were now fair game for the band. The album embraces a host of stylistic influences, including folk, music hall, brass band, circus, vaudeville, avant-garde, Indian, and Western classical music. It was the most expensive album yet produced, and an artwork in itself. For the first time ever in popular music, the songs' lyrics were printed on the album sleeve.

With the ever-evolving recording techniques at their disposal, The Beatles were conjuring a kaleidoscope of sound the likes of which had never been heard before on a pop/rock album. Truly gone were the days of *Please Please Me* and laying down an entire album in a day. The Beatles would spend an unprecedented five months recording *Sgt. Pepper*. Now that they'd abandoned touring, they no longer needed to compose songs which needed to be played live. So, along with the invaluable guidance of George Martin, the band turned Abbey Road into an audio laboratory. Here they pushed the possibilities of recording tech to new boundaries. Together, the five "Beatles" conjured a new sound, instrument by instrument, layer by layer, track by track—a revolutionary method that would soon become the standard in all recording studios the world over. And so, critics lauded *Sgt. Pepper* for its innovations in songwriting, production, and graphic design, and for bridging the divide between popular music and high art. As we shall see, there are innovations and revelations on every track of *Sgt. Pepper*.

Once the album was finished, the story goes that the band left Abbey Road with an acetate copy of the recordings and drove to the pad off the King's Road belonging to "Laurel Canyon's Gertrude

Sgt. Pepper's Lonely Hearts Club Band (1967)

Stein," Mama Cass Elliott of The Mamas & the Papas. There, at six in the morning, they threw open the windows, placed some speakers on a ledge, and blasted the album full pelt over the Mary Poppins rooftops of Chelsea.

"All the windows around us opened and people leaned out, wondering," according to the band's press officer, Derek Taylor. "It was obvious who it was on the record. Nobody complained. A lovely spring morning. People were smiling and giving us the thumbs up."

Two months later, *Sgt. Pepper's Lonely Hearts Club Band* was released. It was a huge cultural happening. The album entranced the old as well as the young. EMI boss Sir Joseph Lockwood reported seeing a group of rich older women at a party so "thrilled" by *Sgt. Pepper* that they all sat on the floor after dinner belting out versions of the songs. In the US, some radio stations suspended their usual playlists for several days in favor of exclusively playing tracks from *Sgt. Pepper*. A quasi-religious reverence greeted the album. Paul Kantner of Jefferson Airplane recalls that David Crosby of The Byrds brought a recording of *Sgt. Pepper* to a hotel in Seattle and played it all night in the lobby. A hundred young music fans sat enraptured on the stairs, as if attending some kind of spiritual experience. In Kantner's words, "Something enveloped the whole world at that time, and it just exploded into a renaissance."

American political theorist Langdon Winner observed that the week *Sgt. Pepper* was released was "the closest Western Civilization has come to unity since the Congress of Vienna in 1815.... At the time I happened to be driving across the country on Interstate 80; in each city where I stopped for gas or food, Laramie, Ogallala, Moline, South Bend, the melodies wafted in from some far-off transistor radio or portable hi-fi. It was the most amazing thing I've ever heard."

Track by Track: *Sgt. Pepper's Lonely Hearts Club Band*

"Sgt. Pepper's Lonely Hearts Club Band"	Recorded: February 1/2, March 3/6, 1967	McCartney *(vocals, bass guitar, lead guitar)*	Lennon *(harmony vocals)*
(Lennon-McCartney)	UK Release: June 1, 1967	Harrison *(harmony vocals, guitar)*	Starr *(drums)*

Additional contributors: Various unnamed French horn players

After touring relentlessly for the past five years or so, the band had decided enough was enough. In November 1966, McCartney and loyal roadie Mal Evans set off on a well-deserved holiday now that they no longer had any touring commitments. It was on the return flight from Kenya to London that the idea of *Sgt. Pepper* began to crystallize in McCartney's mind.

Bored with being the lovable "mop tops," McCartney had the idea of creating a fictional alter-ego band in order to free them from the shackles of being The Beatles and setting their creativity free. Thus, the idea of the *Sgt. Pepper's Lonely Hearts Club Band* was born, a tongue-in-cheek title thought up by McCartney and Evans in the style of contemporary San Francisco-based groups such as Big Brother and the Holding Company.

Written by McCartney, the album's title track, and reprise, starts with the sound of an orchestra tuning up, with the added sound of audience members chatting as they wait for the band's live performance to begin. Fusing rock music with a French horn quintet and a brass ensemble, and punctuated with three-part harmonies from McCartney, Lennon, and Harrison, the song serves to introduce the audience to the *Sgt. Pepper* concept. The main version ends with musical compère McCartney introducing yet another alter ego, Billy Shears.

Sgt. Pepper's Lonely Hearts Club Band (1967)

"With A Little Help From My Friends"	Recorded: March 29/30, 1967	McCartney *(backing vocals, bass guitar, piano)*	Lennon *(backing vocals, cowbell)*
(Lennon-McCartney)	UK Release: June 1, 1967	Harrison *(lead guitar)*	Starr *(vocals, drums)*

Additional contributors: George Martin (Hammond organ)

As conspiracy theories go, this one was a doozy. A ridiculous rumor began circulating that Paul had died in 1966 and was "secretly" replaced by a doppelgänger. According to this theory, McCartney had died in a car crash and, to kindly spare the fans from grief, the surviving Beatles, aided by none other than Britain's security services, replaced Paul with a facsimile. This is no mean feat when you consider McCartney was not only an attractive young man but also became the most commercially successful songwriter in the history of popular music! (Perhaps the conspiracy theorists were influenced by *Thunderball*, the 1965 James Bond movie in which a S.P.E.C.T.R.E. agent undergoes plastic surgery to become a doppelgänger of a real pilot as part of yet another dastardly spy-fi plot.)

Anyhow, the surviving Beatles, including the "new" McCartney couldn't help communicate their own dastardly plot by unnecessarily dropping secret and subtle clues in subsequent albums. You can only imagine the fun the band had with this theory. Indeed, in *The Lyrics*, McCartney himself says of "With A Little Help From My Friends," "Billy Shears was the name of the person who supposedly replaced me in The Beatles when I'd 'died' after a road accident in 1966. That was a crazy rumor that had been doing the rounds. Now Billy Shears showed up, large as life, in the guise of Ringo Starr!"

"We tailored it especially for [Ringo]," McCartney continued, "who had a style of singing different to ours and I think that's one reason why it was such a great success for him on *Sgt. Pepper*."

In typical Beatles fashion, McCartney's explanation of the song's lyrics range from the classical to the vulgar: "'Lend me your ears'—well, you know where *that's* from. The Bard's four hundredth birthday had fallen in April 1964, and there'd been a production of *Julius*

Caesar on television that year. It was still fresh in our minds. John and I were able to include one or two little private jokes here: 'I get high with a little help from my friends . . . What do you see when you turn out the light?' I was imagining turning out the light when you're in bed, under the covers. You're talking about your genitals; that's what it is. Everyone does that: touching themselves when the light goes out. But I couldn't say, 'What do you see when you turn out the light? Your dick.' It just doesn't scan!"

The backing track was recorded in ten takes on March 29, 1967, and on the following day, on the morning of which they posed for that famous *Sgt. Pepper* album cover just off the King's Road in Chelsea, the recording of the rest of the track was completed. Joe Cocker's version of the song topped the UK singles chart in 1968 and, thanks to Cocker's soulful and convulsing performance of the song at Woodstock, "With A Little Help From My Friends" transcended from album track to generational anthem.

"Lucy In The Sky With Diamonds"	Recorded: February 28, March 1/2, 1967	McCartney *(harmony vocals, bass guitar, organ)*	Lennon *(double-tracked vocals, lead guitar)*
(Lennon-McCartney)	UK Release: June 1, 1967	Harrison *(harmony vocals, lead guitar, acoustic guitar, tambura*	Starr *(drums, maracas)*

Although legend claims that Lennon wrote the song as an ode to LSD, of which he was a daily consumer, the man himself claimed the idea came from a more innocent source: a pastel drawing by his four-year-old son Julian.

"I had no idea it spelt LSD. *This is the truth*: my son came home with a drawing and showed me this strange-looking woman flying around. I said, 'What is it?' and he said, 'It's Lucy in the sky with diamonds,' and I thought, 'That's beautiful.' I immediately wrote a song about it."

The song's imagery is clearly inspired by Lewis Carroll's *Alice's Adventures in Wonderland* which Lennon greatly admired, as Julian's

Sgt. Pepper's Lonely Hearts Club Band (1967)

drawing had reminded him of *Through the Looking Glass* in which Alice floats in a "boat beneath a sunny sky." The song was one of the fastest recordings made for the album: one night for the rhythm track and one for the overdubs. It was also the most heavily vari-speeded song on the album.

Musically, the track is driven along by a Lowrey organ played by McCartney, which opens the track, together with a tambura drone played by Harrison that gives the track a wonderful, dreamy timbre. A total psychedelic experience. Incidentally, while some of the drug claims were spurious, it would be daft to pretend that *Sgt. Pepper* wasn't fundamentally shaped by LSD. Many commentators have stated that the album's soundscape, especially its use of echo and reverb, arguably remains the most authentic aural simulation of the psychedelic experience ever made. Moreover, the listener gets the strong feeling that something "other" dwells herein: some kind of distillation of the spirit of the Summer of Love, and it was perceived as such by huge numbers of music lovers who may never have taken LSD.

Sgt. Pepper's Lonely Hearts Club Band—especially tracks like "Lucy In The Sky With Diamonds"— may not have created the psychic atmosphere of the day but, as a simulacrum, this famous album enhanced and radiated the feeling around the globe.

"Getting Better"	Recorded: March 9, 10, 21, 23, 1967	McCartney *(double-tracked vocals, bass guitar)*	Lennon *(backing vocals, lead guitar)*
(Lennon-McCartney)	UK Release: June 1, 1967	Harrison *(backing vocals, lead guitar, tambura)*	Starr *(drums, congas)*

Additional contributors: George Martin (piano)

In 1964, as the band were about to embark on their world tour, Ringo was taken to a hospital in London with suspected tonsillitis and pharyngitis and advised to rest. His replacement for eight days was Jimmy Nichol, who stood in for the gigs in the Netherlands, Hong Kong, and Australia. When asked how he was getting on, Nichol would reply, "It's getting better." Three years later McCartney, with his trademark optimism, remembered the phrase and used it as the title for this song.

Co-written with Lennon, the lyrics reveal the contrasting personalities of the two songwriting partners in a similar way to "We Can Work It Out." As McCartney recalled, "I was just sitting there doing 'Getting better all the time' and John just said in his laconic way, 'It couldn't get no worse,' and I thought, *Oh, brilliant!* This is exactly why I love writing with John."

While the feel of the song is upbeat, Lennon's lyrics are dark and confessional, expressing anger and violence towards women. Musically, the track is lifted by McCartney's wonderful bass part, acting as a melodic counterpart to the droning of Harrison's tambura.

During the *Sgt. Pepper* sessions, McCartney's bass part was recorded after the main track was completed, usually in the early morning hours after the rest of the band had gone home. According to Geoff Emerick, "Being able to work off of all the other elements of the track, including lead and backing vocals, enabled him to hear the song as a whole and therefore create melodic basslines that perfectly complemented the final arrangement." More than ever, McCartney's harmonically intricate basslines were becoming increasingly important to the band's sound.

"Fixing A Hole"	Recorded: February 9/21, 1967	McCartney *(double-tracked vocals, bass guitar, lead guitar)*	Lennon *(backing vocals)*
(Lennon-McCartney)	UK Release: June 1, 1967	Harrison *(backing vocals, double-tracked lead guitar)*	Starr *(drums, maracas)*

Additional contributors: George Martin (harpsichord)

In Greek mythology, the Muses were the nine goddesses who served as the sources of inspiration for artists, poets, musicians, and thinkers. Artists would invoke the Muses to seek their blessings and

Sgt. Pepper's Lonely Hearts Club Band (1967)

inspiration, believing that they would bestow creative genius upon them. Compared to this, McCartney's miracles seem godless.

As he says in *The Lyrics*, "Before I write a song, there's a black hole and then I get my guitar or piano and fill it in. The notion that there is a gap to fill is no less honorable a basis for an inspiration than a bolt of lightning coming down out of the sky. One way or another, it's a miracle."

The inspiration for "Fixing A Hole" was LSD. McCartney said again in *The Lyrics*, "I was the last in the group to take LSD . . . I was very reluctant because I'm actually quite straitlaced. . . . In the end I did give in and take LSD one night with John. . . . Around that time, when I closed my eyes, instead of there being blackness there was a little blue hole. It was as if something needed patching. I always had the feeling that if I could go up to it and look through, there would be an answer. . . . The fact is that the most important influence here was not even the metaphysical idea of a hole, which I mentioned earlier, but this absolutely physical phenomenon—something that first appeared after I took acid. I still see it occasionally, and I know exactly what it is. I know exactly what size it is."

Moreover, in *Many Years From Now*, McCartney explained that "mending was my meaning. Wanting to be free enough to let my mind wander, let myself be artistic, let myself not sneer at avant-garde things."

Much later, for example, McCartney tried to put an experimental track on *Anthology 2*, but George Harrison voted to reject it. As McCartney explained in October 2021 on BBC Radio 4's *This Cultural Life*, Harrison "didn't like avant-garde music" and referred to avant-garde as "avant-garde a clue" ("haven't got a clue!"). Some of the recording sessions for "Fixing A Hole" were at Regent Sound Studios in London, as Abbey Road Studios were unavailable.

Also present at one session was Jesus Christ. Or at least a young man who thought himself to be Christ. He had arrived at McCartney's house in St. John's Wood, just before McCartney was due to depart for Regent Sound. As McCartney later recalled, "There were a lot of casualties about then. We used to get a lot of people who were maybe insecure or going through emotional breakdowns or whatever. So I said, 'I've got to go to a session but if you promise to be very quiet and just sit in a corner, you can come.' So he did, he came to the session and he did sit very quietly and I never saw him after that."

"She's Leaving Home"	Recorded: March 29/30, 1967	McCartney *(double-tracked vocals, backing vocals)*	Lennon *(double-tracked vocals, backing vocals)*
(Lennon-McCartney)	UK Release: June 1, 1967		

Additional contributors: Various unnamed players on violins, violas, cellos, double bass, and harp

"She's Leaving Home" was inspired by a story in a British newspaper about a teenage girl who had run away from home. As McCartney explains in *The Lyrics*, "This one is based somewhat on a newspaper report of a missing girl. The headline was something like 'A-Level Girl Dumps Car and Vanishes.' So I set out to imagine what might have happened, the sequence of events."

His other inspiration came from seminal British film director Ken Loach, "In addition to the newspaper report, another influence was *The Wednesday Play*. It was a weekly television play that often addressed 'big' social issues. One of the most famous of these plays was *Cathy Come Home*, directed by Ken Loach." McCartney and Lennon jointly wrote the lyrics, with McCartney writing the main narrative in the verse while Lennon contributed the anguished views of the girl's parents in the chorus.

Just like "Eleanor Rigby" the year before, the recording does not feature any instrument played by a Beatle; only the vocals of McCartney and Lennon are present. The instrumental backing of the track was played by an orchestra arranged by Mike Leander: harp, violins, violas, cellos, and a double bass. George Martin was not able to score the backing track as he was busy producing with Liverpudlian

Sgt. Pepper's Lonely Hearts Club Band (1967)

singer Cilla Black and was hurt by McCartney's impatience in hiring Leander at short notice. Lennon and McCartney received the 1967 Ivor Novello award for "Best Song Musically and Lyrically" for this song.

"Being For The Benefit Of Mr. Kite!"	Recorded: February 17/20, March 28/29/31, 1967	McCartney *(bass guitar)*	Lennon *(double-tracked vocals, organ)*
(Lennon-McCartney)	UK Release: June 1, 1967	Harrison *(harmonica)*	Starr *(drums, tambourine, harmonica)*

Additional contributors: George Martin (various instruments)

On the last day of January in 1967, the band had been in Sevenoaks in Kent, filming a promo for "Strawberry Fields Forever," about which more later. While on location, Lennon had sauntered into an antiques shop and picked up a most curious Victorian circus poster. It boasted "the last night but three" (!) of a show staged by a troupe of traveling tumblers in Rochdale on Valentine's Day in 1843. The poster appealed hugely to Lennon's fine sense of the quirky and, when the production schedule of *Sgt. Pepper* called for another contribution from him, Lennon fixed the flyer on the wall of his home studio and, playing his piano, simply sang snatches from it until he had a song.

Lennon later claimed to still have the poster at his home, and that practically "everything from the song is from that poster, except the horse wasn't called Henry." (The horse of the poster is "Zanthus"!) This seemingly unimportant detail led to "Being For The Benefit Of Mr. Kite!" to becoming one of three songs from *Sgt. Pepper* that was banned from being broadcast on the BBC, ostensibly because the phrase "Henry the Horse" was a fusion of two words that were each known as slang for heroin. Understandably, given its provenance, Lennon denied the song had anything to do with heroin.

Taking "Mr. Kite!" to Abbey Road, Lennon asked George Martin for a "fairground" production so strong that one might "smell the sawdust." What they then achieve together is something akin to musical time travel. At the same time as being the album's most poignant tribute to a bygone era, "Mr. Kite" is also one of the most technologically challenging tracks on *Sgt. Pepper*. Recording resources were still very primitive in 1967. In our digital present, an artist or producer can create as many layers of sound as they wish, with potentially hundreds of separate recordings all piled onto the same aural layer cake.

The only recipe available to George Martin and the band at the time was the proverbial four separate tracks on a single magnetic tape. So they had to be very creative in their solutions. A popular tactic they adopted was to use up all four tracks on tape machine number one (vocals, bass, drums, and harmonium), making a mix of those four. They would then record that mix, or "bounce" it, onto a single track of the four tracks available on tape machine number two. Cunning.

Indeed, this is exactly what they did for "Mr. Kite." They laid down a mix of Lennon's guide vocal, Paul's bass, Ringo's drums, and George Martin's harmonium as the four tracks on tape machine number one. Apart from the guide vocal, which was dumped, these instruments made the final mix, which was bounced down to a single track of tape machine number two. And so, the process was repeated. Why wasn't everything just recorded at once? Simply because these aural creations were far too complex for that. And some layers simply couldn't be laid down by live performance.

For instance, consider the part of the song where the BBC-banned "Henry the Horse" dances "the waltz," which is roughly one full minute into the track. What you hear here is the result of Lennon's instruction wishing George Martin would create the atmosphere of an old-fashioned circus purely in musical terms. *You are smelling the sawdust*, as it were. Martin decided the best results would be achieved by using the sounds of a fairground organ. But these *calliopes*, as they are called, are not keyboards. They create sound through the sending of gas, originally steam then, more lately, compressed air, through large whistles, or by the operation of punch cards like a player piano.

So, Martin gathered together as many pre-recorded tapes of calliopes as he could unearth, and asked engineer Geoff Emerick to chop them up into one-second snippets, throw them in the air, then randomly stick the snippets together again in a kind of chaotic and cacophonous ticker-tape of sound. The result is stunning and quite brilliant. The revolutionary sound production on this song is one of the very first examples of "sound sampling." The overall effect on the listener is transformational; we are treated to an archly comic and surreal soundtrack, a time travel movie with sounds of our circus past, but one which a real calliope could never have conjured up.

| "Within You Without You" | Recorded: March 15/22, April 3/4, 1967 | | |
| (Lennon-McCartney) | UK Release: June 1, 1967 | Harrison *(vocals, sitar, acoustic guitar, tambura)* | |

Additional contributors: Various unnamed players on violins, cellos, tambura, tabla, dilrubas, svarmandal

Some people playing *Sgt. Pepper* for the first time, when flipping the vinyl over to "side two," might have been forgiven for thinking there was something seriously wrong with their pressing. Dropping the stylus onto George Harrison's "Within You Without You," they might have thought they'd teed up the wrong record. The listener had already been treated to a turn-of-twentieth-century bandstand with the title track, "Sgt. Pepper's Lonely Hearts Club Band," the *Alice in Wonderland* phantasmagoria of "Lucy In The Sky With Diamonds," the north of England kitchen sink drama that is "She's Leaving Home," and the Victorian circus sounds of "Being For The Benefit of Mr. Kite!" And now, this.

The listener is taken on a trip to another continent completely. One's senses are at once subjected to the sounds of the sitar, the dilrubas, the svarmandal, and the tabla; all mixed in with a heady concoction of cellos and violins. "Within You Without You" is the most esoteric track on *Sgt. Pepper,* and arguably the most visionary of the entire collection. It is safe to say that this track is a tipping point in twentieth century popular music, and acted as an inspiration for what we now know as "world music." The revolutionary nature of this song isn't just down to the mere use of Indian instruments. The likes of The Beatles and The Rolling Stones had used them before in their music to add an exotic feel and flavoring to their compositions. *Within You Without You* went a step further. The song embraces the very tenets of Indian music. Back in 1965, George had first come across a recording of the great virtuoso of classical Indian music, Ravi Shankar. It was an epiphany in Harrison's musical evolution.

By 1966, George was taking lessons from Shankar. The great religion of Hinduism, which entails a whole host of diverse systems of thought and embraces concepts that discuss theology, mythology, and many others, has its chief focus on the hazards and illusions of the material world. Increasingly, in the west in the 1960s, people were welcoming Hinduism as part and parcel of a new spirituality. For many, it merely proved a passing fashion. But for Harrison, Hinduism became an earnest and abiding adherence. George studied Hindu religious philosophy as well as Indian music intently. And it was this amalgam of Indian music and philosophy that George was ambitiously trying to express in a pop album.

In short, "Within You Without You" is a state-of-being-in-the-cosmos report on the modern world in 1967. As George says at the end of the song, "When you've seen beyond yourself / Then you may find peace of mind is waiting there / And the time will come when you see we're all one / And life flows on within you and without you."

To carry off this radical summation of Indian music and philosophy in one song, George had to grasp the crucial distance between Indian and Western music. In Indian music, rhythmic patterns play a far more important role in how a song is constructed. In Western music, typically, the rhythm won't evolve during a song. But in Indian music, rhythm rarely stays the same for any length of time. Thus, "Within You Without You" sounds irregular to the untrained Western ear and upsets the listener's expectation as to where the strong beats will fall. And the abstruse rhythms are unheard of, even in Indian music; they are entirely of George's making.

Thrown by the track's apparent lack of harmonic interest, Harrison's contribution to *Sgt. Pepper* didn't fare well with many listeners. George's lyrics didn't help. Pointing out how trivial the Me Generation is was never the best plan for profit. And yet, there was always something paradoxically misanthropic about George's songs. From "Don't Bother Me" (what a start!) to the infamous "Piggies" (more on this later), Harrison's compositions have an air of superiority and overly pious finger-wagging about them. Those "who gain the world and lose their soul."

As to the laughter at the end of the track, American musicologist Alan Pollack suggests there are two schools of thought on the matter. The first is that the laughter comes from the *Sgt. Pepper* audience within the fiction of the album's conceptual world.

"The xenophobic audience ([with their] Victorian/Edwardian-era outlook of supercilious Imperialism)," Pollack posits, "is letting off a little tension of this perceived confrontation with pagan elements."

And the second school of thought is "the bedazzled composer, in an endearingly sincere nanosecond of acknowledgment of the apparent existential absurdity of the son-of-a-Liverpudlian bus driver espousing such otherworldly beliefs and sentiments, is letting off a bit of his own self-deprecating steam in reaction to the level of true courage expended by him in order to come out of the uneasily-anti-materialistic closet." Phew.

"When I'm Sixty-Four"	Recorded: December 6/8/20/21, 1966	McCartney *(vocals, bass guitar, piano)*	Lennon *(backing vocals, guitar)*
(Lennon-McCartney)	UK Release: June 1, 1967	Harrison *(backing vocals)*	Starr *(drums, bells)*

In the writing of "When I'm Sixty-Four," McCartney identifies in *The Lyrics* his debt to Irish poet, Louis MacNeice:

One influence was the humor of Louis MacNeice's poem "Bagpipe Music:" "John MacDonald found a corpse/Put it under the sofa/Waited till it came to life/And hit it with a poker/Sold its eyes for souvenirs/Sold its blood for whisky/Kept its bones for dumb-bells to use when he was fifty." MacNeice is great on the day-to-day. I think he would recognize, "You can knit a sweater by the fireside/Sunday mornings, go for a ride." All comfortable things that retired people do. Then I would stick in "Doing the garden, digging the weeds." "Digging the weed" is also a way of saying "enjoying a little pot." We would always slip in those little jokes because we knew our friends would get them.

From "Penny Lane" and "Strawberry Fields Forever," an informal organizing theme of these recording sessions had been the band's upbringing in Liverpool. These evocations of childhood include the circus and fairground effects, the pervading atmosphere of traditional northern English music hall, and Lewis Carroll-inspired imagery, which acknowledges John's favorite childhood reading. Indeed, in a 1995 interview, McCartney recalled that the Liverpool childhood theme served as a device or underlying theme throughout the *Sgt. Pepper* project.

McCartney had the melody for "When I'm Sixty-Four" "fully worked out" by the time he was about sixteen. And, given the melody had "something of a music hall feel," it was perfect for *Sgt. Pepper.*

The song was originally recorded in a different key from the final version, then sped up, as McCartney had asked that his voice sound younger. The final production prominently features a trio of clarinets. George Martin recalled that he remembered recording the track "in the cavernous Number One studio at Abbey Road and thinking how the three clarinet players looked as lost as a referee and two linesmen alone in the middle of Wembley Stadium."

Engineer Geoff Emerick later explained, "The clarinets on that track became a very personal sound for me; I recorded them so far forward that they became one of the main focal points." Meanwhile, British comic book author Alan Moore (he of *Watchmen*, *V for Vendetta*, and *From Hell*) views "When I'm Sixty-Four" as a fusion of ragtime and pop, demonstrating the diversity of *Sgt. Pepper* and suggesting that the music hall atmosphere is reinforced both by McCartney's vocal delivery and the recording's use of chromaticism, a harmonic pattern that can be traced to Scott Joplin's "The Ragtime Dance" and Johann Strauss' "The Blue Danube."

"Lovely Rita"	Recorded: February 23/24, March 7/21, 1967	McCartney *(vocals, bass guitar, piano, comb and paper*	Lennon *(backing vocals, vocal percussion, acoustic rhythm guitar, comb and paper)*
(Lennon-McCartney)	UK Release: June 1, 1967	Harrison *(backing vocals, acoustic rhythm guitar, comb and paper)*	Starr *(drums, comb and paper)*

Additional contributors: George Martin (piano)

"Lovely Rita" serves as McCartney's affectionate ode to a traffic warden (or "meter maid" as they are called in the US.) Its composer was amused by the idea of falling in love with such a disliked "authority figure":

> Nobody liked parking attendants, or meter maids, as they were known in that benighted era. So, to write a song about being in love with a meter maid—someone nobody else liked—was amusing in itself. There was one particular meter maid in Portland Place on whom I based Rita. She was slightly military-looking.

The majority of the lyrics were written by McCartney with a little help from Lennon during the recording session. Once the main rhythm track was recorded, work began on the vocals. McCartney was looking for a Beach Boys–style vocal arrangement. The vocals themselves were recorded with heavy tape echo, especially on the backing vocals led by Lennon.

"John always wanted repeat echo in his headphones, it gave him more excitement," says Geoff Emerick.

To enhance the emotional impact of Rita, Lennon added some heavy breathing, sighing, and screaming noises for the end of the song. Sound effects added consisted mainly of the percussive sound of the four Beatles standing around a single microphone humming through hair combs wrapped with Abbey Road's lavatory paper stamped with the words "Property of EMI." To finish off the track, George Martin added a honky-tonk style piano solo recorded with the usual varispeed effect and with heaps of tape echo.

"Good Morning Good Morning"	Recorded: February 8/16, March 13/28/29, 1967	McCartney *(backing vocals, lead guitar, bass guitar)*	Lennon *(double-tracked vocal, rhythm guitar)*
(Lennon-McCartney)	UK Release: June 1, 1967	Harrison *(backing vocal, lead guitar)*	Starr *(drums, tambourine)*

Additional contributors: Various unnamed players on saxes, trombones, and French horn

Inspired by a Kellogg's cereal television commercial, Lennon's "Good Morning Good Morning" critiques the banality of suburban life.

"John was feeling trapped in suburbia," according to McCartney. "[He] was going through some problems with Cynthia. It was about his boring life at the time; there's a reference in the lyrics to 'nothing to do' and 'meet the wife' . . . he was that bored, but I think he was also starting to get alarm bells."

This uneasiness of being trapped is mirrored in the shifting time signatures heard throughout the song. Starting with a cock crowing, the song works its way through a plethora of animal sound effects, courtesy of EMI's sound effects tapes, finally ending with the clucking

Sgt. Pepper's Lonely Hearts Club Band (1967)

of a hen that segues nicely into the opening guitar screech of the next song, "Sgt. Pepper's Lonely Hearts Club Band (Reprise)."

The seemingly random animal sound effects were in fact arranged in specific order.

"The idea was that, as the music was fading away, the sounds of various animals would be heard, with each successive animal capable of chasing or frightening the next animal in line. John had actually thought this through to the extent that he'd written down a list of the animals he wanted on there, in order."

Musically, the song is underpinned by the stomping brass arrangement courtesy of Sounds Incorporated, old Liverpool friends of the band going back to their days in Hamburg, with a searing Indian-inspired guitar solo by McCartney. Given all this, it's unsurprising that the song ended up with the distinction of being the *Pepper* track with the largest number of overdubs.

"Sgt. Pepper's Lonely Hearts Club Band (Reprise)"	Recorded: April 1, 1967	McCartney *(vocals, bass guitar, organ)*	Lennon *(vocals, rhythm guitar)*
(Lennon-McCartney)	UK Release: June 1, 1967	Harrison *(vocals, lead guitar)*	Starr *(vocals, drums, tambourine, maracas)*

As if to bookend the title track, "Sgt. Pepper's Lonely Hearts Club Band" is reprised here towards the end of the album, albeit at a faster tempo and with a much heavier sound of distorted guitars. The idea for this reprise came from the band's road manager, Neil Aspinall, who reasoned that, given there was a "welcome song" at the start of proceedings, there should be a "goodbye song" too.

"A Day In The Life"	Recorded: January 19/20, February 3/10/22, 1967	McCartney *(vocals, bass guitar, piano)*	Lennon *(double-tracked vocals, acoustic guitar, piano)*
(Lennon-McCartney)	UK Release: June 1, 1967	Harrison *(congas)*	Starr *(drums, maracas)*

Additional contributors: Various players in orchestration

Reaction channels are one of the most popular kinds of content on today's YouTube and Twitch. Users seem to delight in the entertainment of watching a channel's host react to different aspects of culture; be it films, TV programs, news, movie trailers, or music. And that's just as well when it comes to "A Day In The Life."

Listening to the track for the first time, you might as well have company of some kind, as when that terminal piano chord fades out, you may find yourself gazing out your window to see if reality is still what you thought it was. After all, before the band embarked on the recording sessions, Lennon told George Martin the song should "sound like the end of the world." And it does. Just one single reaction-channel example; at the time of writing and with over a hundred-thousand views, a young American drummer reacting on his YouTube channel (*L33Reacts*) to the hugely ambitious climax of "A Day In The Life" (apart from the inevitable reaction of "wtf, dude") says "Bro, I've just had a life-changing vision while listening to this song. How have I been missing this? The *balls* on these guys."

To be fair, by early 1967, the band reigned supreme. They'd just recorded "Penny Lane" and "Strawberry Fields Forever"; the greatest double-A side in popular music and the greatest single coupling in history. "A Day In The Life" came next.

The song embraces the everyday realism of ordinary life. It features a rather deadpan commentary fused with psychedelia, first person and third person narratives juxtaposed, creative use of classical orchestration, use of sound effects, and, naturally, avant-garde techniques and the band's usual experimentation with recording technology. Buried like a *matryoshka* within its album mother-shell, "A Day In The Life" is *Sgt. Pepper* in microcosm.

In January of 1967, two news reports in the *Daily Mail* had caught Lennon's inquisitive eye. One was the inquest on Irish socialite Tara Browne, a twenty-one-year-old male acquaintance who'd been killed in a car crash. (On December 18 in 1966, Browne, a connoisseur of the London counterculture and, like many of its members, a dabbler in "mind-enhancing" drugs, drove his sky-blue Lotus Elan at high speed through red lights in Kensington, ploughing into a parked van and killing himself.) The other was a short news item noting there were four thousand potholes in the roads of Blackburn; to quote the actual report: "There are four thousand holes in the road in Blackburn, Lancashire, or one twenty-sixth of a hole per person, according to a council survey."

Between these two tales, Lennon put in a verse in which his jaded narrator plays onlooker as the English army wins a war. Perhaps prompted by Lennon's acclaimed role in the movie *How I Won the War* three months before; this may have been a subtle reference to Vietnam which, though a real issue to Lennon, might have over-baked the song if commented upon more crudely.

Rather than intellectualizing these reports on "la mort" and the mundane, Lennon riffed on the estrangement and began to write. That same night, he called at McCartney's house with his new composition. Done and dusted, most of the lyrics were set to a trio of verses and an accompanying tune. Now, knee-to-knee, like back in the Liverpool of their teenage days, Lennon and McCartney drilled down to the details. Risqué lines like "I'd love to turn you on" were conjured up, as was the dreamy middle eight; McCartney's account of a kid rushing to school: "Woke up, fell out of bed/Dragged a comb across my head . . . /Found my coat and grabbed my hat/Made the bus in seconds flat/Found my way upstairs and had a smoke/And somebody spoke and I went into a dream."

A couple of days later they began recording the track; three typical band sessions, then two studio aural and psychedelic "happenings" from which the final track derives its immense power. In its early stage, the track's composition was so new that they hadn't time to work out how to join John's section to Paul's, so they simply recorded a basic track and left a few random gaps of twenty-four bars (the gaps are counted in by road manager Mal Evans, whose voice was left on the final recording.) It was Paul who then decided that these gaps be filled by the orchestral climax.

McCartney had been listening to avant-garde music by John Cage and Luciano Berio and decided the twenty-four bar bridges would be best packed by an orchestra running from its lowest to its highest note in an unsynchronized slide: a kind of musical hippie "freak-out." These hippie events became the McCartney-inspired Abbey Road parties where tuxedoed classical orchestra musicians wore red noses and gorilla paws as they readied to create one of music's most stirring and cacophonous climaxes. It's quite sublime that "A Day In The Life" starts so inconspicuously, crossfading out of the frantic "Sgt. Pepper's Lonely Hearts Club Band" reprise and ultimately ends in cacophony.

George Martin and McCartney marshaled the musicians to create this track's climax. As Martin explains in Mark Lewisohn's *The Complete Beatles Recording Sessions*:

> At the very beginning I put into the musical score the lowest note each instrument could play, ending with an E major chord. And at the beginning of each of the 24 bars I put a note showing roughly where they should be at that point. Then I had to instruct them. "We're going to start very very quietly and end up very very loud. We're to start very low in pitch and end up very high. You've got to make your own way up there, as slidey as possible so that the clarinets slurp, trombones gliss, violins slide without fingering any notes. And whatever you do, don't listen to the fellow next to you because I don't want you to be doing the same thing." Of course they all looked at me as though I was mad ...

At the end of this festive happening, the "partygoers" left in the studio spontaneously applauded the musical outcome. That most final of final chords on "A Day In The Life"—played by John, Paul, Ringo, George Martin, and Mal Evans on three pianos (and tracked four times)—wasn't recorded until a dozen days later. The whole thing was completed in around thirty-four hours, total.

And so, "A Day In The Life" plays out on a vast and open canvas. The start of the song is deceptively simple, albeit laced with Lennon's laconic descending harmonies. But soon, the gravitas of the opening chords on an acoustic guitar begins to build our musical journey towards that crushing and crescendo'd finale. Lennon's vocal decision is deliberate. As the first ominous piano chords rise up, he sings, "I read the news today, oh boy," and his delivery feels like that of the detached reporter, world-weary and meditative without being proud. Below Lennon's deadened vocal, energy groundswells from the bass and pianos until, at the cue of "he blew his mind out in a car," Ringo's complex and syncopated tom-toms start tattooing their response. Ringo's drums reject the traditional rock/pop role of percussive rhythm of the day in favor of a more classical approach, whereby his slack-tuned tom-tom fills punctuate the pulse of the track, disorients the listener, and the musical landscape shifts once more.

As one of the most influential and celebrated songs in popular music history, appearing on many lists of the greatest songs of all time, and being regularly cited as The Beatles' best song, "A Day In The Life" has had many different interpretations of meaning, with more claptrap penned about this track than anything else the band recorded. Almost single-handedly, this track alone, for many people, answers the question why this album is considered by many musicians to be such a landmark.

The song has variously been interpreted to be a sober return to reality after the intoxicated fantasy of "Pepper-land"; an abstract statement about the structure of the pop album; an account of a bad acid trip; a funereal festival of death; and even pop's version of "The Waste Land," T. S. Eliot's celebrated 1922 poem about brokenness and loss, and which also features abrupt and unannounced changes of narrator, location, and time. Many of these misinterpretations seem oblivious to the fact that, though "A Day In The Life" has the status of the emotional and artistic climax of *Sgt. Pepper* and seems to reflect on many of the album's themes, it was actually the first track the band recorded for the album.

As a self-confessed Beatles obsessive, Oscar-winning actor Cillian Murphy considers "A Day In The Life" the apotheosis of Lennon and McCartney's collaboration.

"The Beatles are kind of like my musical touchstone," Murphy said on BBC Radio 4's *Desert Island Discs* in February of 2024. "I think they probably represent, in my mind, one of the greatest artistic achievements of the twentieth century. Not just musically but in terms of their humor, their friendship, and their tolerance. I love their energy, Paul's hope and optimism and then John's kind of acerbic [realism]. It's just perfection." As Cillian told KCRW radio back in September 2023, "It's very hard to pick your favorite Beatles tune. It's almost impossible. But if I had to pick one, it would be 'A Day In The Life' ... I think it's one of the greatest works of art. I think if you didn't have 'A Day In The Life' you wouldn't have 'Bohemian Rhapsody.' You wouldn't have 'Paranoid Android.' They were just pushing the envelope, you know? And the thing that always gets me about that piece of music is, they were in their mid-twenties. I think George was twenty-four. So, it's just phenomenal. They've been a constant companion to me all through my life. I started listening to them when I was four or five and I've never stopped ... ['A Day in the Life'] ... almost like gave permission for music to become the way it did. Every time I listen to [it], it still moves me. That incredible orchestral orgasm in between Lennon's bit and McCartney's bit, and then that final chord at the end, there's nothing like it really. It's just profoundly brilliant."

For many music critics, "A Day In The Life" is the apogee of the band's artistic achievement. With some of their most disciplined and

Sgt. Pepper's Lonely Hearts Club Band (1967)

persuasive lyrics, its musical expression is spectacular, its structure utterly unique and stunning, and its execution peerless. Lennon's ethereal vocal sits perfectly alongside McCartney's pragmatic realism, as Ringo's drums hold the track together, and George Martin's team of technicians, in working conditions that would no-doubt floor the vast majority of those working in modern studios, amass all the elements of a song which remains among the most powerful and creative artistic reflections of its time.

MAGICAL MYSTERY TOUR (1967)
Experimental Rock

"This free-form associative tinkering happened a lot after Sgt. Pepper, *on Magical Mystery Tour. It was a side to The Beatles that I found rather tedious. I used to say to them, 'If you want to be random, let's be organized about it,' which was definitely not what they wanted to hear when they were in that mood. [. . .] When John brought along* I Am The Walrus *later in 1967, I said, 'I see what you're trying to get out: it's very bizarre, but it's great. Let's organize it.' John went along with that."*

—George Martin, *All You Need Is Ears* (1994)

Magical Mystery Tour	Released: November 27, 1967	Recorded: April 25 – May 3, August 22 – November 7, 1967	Duration: 36:35
Producer: George Martin	Studio: EMI Chappell, London	Label: Parlophone/Capitol	Tracks: 11

Track Listing

Side One

No.	Title	Lead Vocals	Length
1	"Magical Mystery Tour"	McCartney	2:48
2	"The Fool On The Hill"	McCartney	2:59
3	"Flying"	Instrumental	2:16
4	"Blue Jay Way"	Harrison	3:54
5	"Your Mother Should Know"	McCartney	2:33
6	"I Am The Walrus"	Lennon	4:35

Side Two

| 7 | "Hello, Goodbye" | McCartney | 3:24 |
| 8 | "Strawberry Fields Forever" | Lennon | 4:05 |

(Continued)

Magical Mystery Tour (1967)

9	"Penny Lane"	McCartney	3:00
10	"Baby, You're A Rich Man"	Lennon	3:07
11	"All You Need Is Love"	Lennon	3:57

All songs written by Lennon-McCartney except track 4, written by Harrison, and track 3, written by Harrison-Lennon-McCartney-Starkey

Experimental Rock

From the get-go, rock music was innovative and experimental. Originating as "rock and roll" in America during the late 1940s and early 1950s, it owed its provenance and evolution to a creative amalgam of the influences of blues, R&B, and country music, as well as drawing from genres like electric blues, folk, jazz, and other musical styles. However, it wasn't until the late 1960s that rock musicians started to create more complex compositions through advancements in multitrack recording. In the mid-1960s, the boundaries between rock and pop and the avant-garde began to blur as albums were conceived as distinct and artistic statements.

In 1966, Lennon had famously said, in an interview with the *London Evening Standard*, that The Beatles were "more popular than Jesus now; I don't know which will go first—rock 'n' roll or Christianity."

The comment was made and published in a profile of Lennon by Maureen Cleave. This offhanded comment was greeted with considerable hysteria in the US, where many Christians were so outraged that they publicly burned Beatles records in the way Nazis used to burn books. Indeed, the backlash was so bad that it became one of the deciding factors that led to the band's decision to quit touring and focus instead on recording music.

While Lennon's ironic comment on the group's messianic status was arguable before *Revolver*, it seemed pretty realistic after it. *Revolver's* soundscape seemed so revolutionary that Western youth began believing the band were somehow able to divine contemporary culture and orchestrate its essence through their albums. So, the anticipation of the sequel to *Revolver* was intense, to both the general public and among the band's fanbase. From 1965 on, the major rock and pop acts on both sides of the Atlantic had been waging a friendly cultural war to create the most extraordinary music.

By 1967, this had resulted in a huge structural change to the music business. The fleeting turnover of mass-produced and exploited "artistes" melted away to be superseded by a more stable scene, one based on autonomous "artists" no longer manipulated by record company executives. "Artiste" was a more specific term than "artist," mostly referring to professional performers on stage, like a singer, while "artist" can encompass anyone who creates art in any medium, including writers and musicians. (In this context, one imagines the amusing satire of an executive in Pink Floyd's "Have A Cigar": "Well, I've always had a deep respect and I mean that most sincere / The band is just fantastic, that is really what I think / Oh, by the way, which one's Pink?")

This market revolution was easily quantifiable. The turnover of expendable pop acts plunged in inverse proportion to the burgeoning prestige and success of the new rock aristocracy. In short, rather than pushing pap for profit, the new aristocracy were intent on cataloging their times, a mission which meant that the already exciting music scene became consequently thrilling. How times have changed.

Naturally, The Beatles played a major part in this evolution. Historian David Simonelli suggests that, with recordings such as "Tomorrow Never Knows," "Strawberry Fields Forever," "Penny Lane," and "Sgt. Pepper's Lonely Hearts Club Band," the band were established as the most avant-garde [rock] composers of the postwar era. In particular, *Sgt. Pepper* inspired many to believe that experimental rock was commercially viable music. Record executives now lost confidence in what would sell, and money took a backseat in the making of recorded music.

Toward the end of the *Sgt. Pepper* sessions, McCartney began looking for a new project for the band, and on a theater tour of America in

Magical Mystery Tour (1967)

April 1967, he was taken by the zeal of the West Coast hippies compared with their glibber British counterparts. In particular, McCartney was impressed by the Merry Band of Pranksters, the LSD-apostles of American novelist and countercultural figure Ken Kesey. The Merry Pranksters took a Day-Glo-painted International Harvester school bus on tour of the States, filming their counterculture adventures as they went. Himself a cinephile and a maker of underground home-movies, McCartney considered Kesey's psychedelic roadshow as the foundation for a film, which he drafted up while traveling back to the UK. The final mixing for *Sgt. Pepper* was still ongoing when he got back and, rather than let the rest of the band have a few weeks repose, he pushed on with this new project. And so it was, a mere four days after *Sgt. Pepper* was finally in the can, The Beatles were back in Abbey Road, taping the title track for *Magical Mystery Tour*.

Track by Track: *Magical Mystery Tour*

"Magical Mystery Tour"	Recorded: April 25/26/27, May 3, 1967	McCartney *(vocals, bass guitar, piano)*	Lennon *(harmony vocals, acoustic rhythm guitar)*
(Lennon-McCartney)	US Release: November 27, 1967	Harrison *(harmony vocals, lead guitar)*	Starr *(drums, tambourine)*

Additional contributors: Various trumpet players

The blurring of boundaries between rock and the avant-garde found its way into the *Magical Mystery Tour* movie project. Since their touring days had come to an end, the band had eased into a simple routine of ongoing low-intensity recording. This regime suited the hippie times they were living in. As they paid no studio fees, all they needed to do was book a slot at Abbey Road and simply show up, regardless of whether they had anything to tape.

Much of the time they now spent in the studio was concerned mostly with writing songs, before they even got to the rehearsing and recording. Coupled with the new experimental spirit, and their easy schedule, was their increased use of drugs, which now began to influence their judgment and encourage a more relaxed attitude to their work. This title track acts as an overture to both the album and movie, and was written and recorded to be an upbeat and stylized pomp-and-circumstance kind of piece. McCartney began the process with just a few chords and the initial line of the lyric. Developing the track and practicing as the music progressed, he kept the other band members in check by asking them to yell out phrases and batter any percussion on hand, later asking George Martin for traffic sound effects and trumpet overdubs.

When first shown on UK television, on Boxing Day 1967, the movie got a critical panning. And yet, as an early prototype of the road movie sub-genre, it deserves a modest place in film history, with its satire on consumerism, showbiz, and media tropes and clichés. It was The Beatles' take on a counterculture view of bourgeois society.

"The Fool On The Hill"	Recorded: September 25/26/27; October 20, 1967	McCartney *(vocals, bass guitar, piano, acoustic guitar, recorder)*	Lennon *(harmonica, Jaw's harp)*
(Lennon-McCartney)	US Release: November 27, 1967	Harrison *(acoustic guitar, harmonica)*	Starr *(drums, maracas, finger cymbals)*

Additional contributors: Various trumpet players

At the time, Maharishi Mahesh Yogi was the band's spiritual advisor. And "The Fool On The Hill" was written about the time of the band's involvement with the Maharishi. They sought his counsel because The Beatles felt they needed to "re-center" themselves; "to get back to basics," as McCartney puts it.

And he explains that, in the song, he's "simply describing how the Maharishi was perceived by so many people—as the 'giggling guru.' That was not my own perception. I'm fascinated by how much trouble people have in recognizing irony."

And so, McCartney suggests that "The Fool On The Hill" is a "very complimentary portrait and represents the Maharishi as having the capacity to keep perfectly still in the midst of the hurly-burly," and that "he may be related somehow to the truth-telling Fool in *King Lear*."

Described by Alan Pollack as an Early Romantic "art song," "The Fool On The Hill" explores a similar theme of "lonely, alienated isolation" to "Eleanor Rigby" and "She's Leaving Home," with the attention focused almost exclusively, and appropriately, to the main character's inner life. The song is also a fitting exemplar of the way in which The Beatles garnered a unique reputation for rich and extravagant production values at this time, from the years 1967 into 1968.

"Flying"	Recorded: September 8/28, 1967	McCartney *(vocals, bass guitar, guitar)*	Lennon *(vocals, organ, Mellotron)*
(Harrison-Lennon-McCartney-Starkey)	US Release: November 27, 1967	Harrison *(vocals, guitar)*	Starr *(vocals, drums, maracas)*

The first Beatles track to credit all four band members as songwriters, "Flying" was recorded as incidental music for the *Magical Mystery Tour* movie. The song holds the distinction of being the only Beatles instrumental to be released by EMI. A simple twelve-bar blues with a melody written by McCartney and played by Lennon on a varispeeded Mellotron, all four Beatles added vocals described in Mark Lewisohn's *The Complete Beatles Recording Sessions* as "scat chanting." Ringo is credited formally by his birth name of Starkey.

"Blue Jay Way"	Recorded: September 6/7, October 6, 1967	McCartney *(backing vocals, bass)*	Lennon *(backing vocals)*
(Harrison)	US Release: November 27, 1967	Harrison *(double-tracked vocals, backing vocals, Hammond organ)*	Starr *(drums, tambourine)*

Harrison's only songwriting contribution to *Magical Mystery Tour*, "Blue Jay Way" was written in the fog-bound Hollywood Hills of

California while waiting for the band's publicist, Derek Taylor, to arrive. In order to alleviate his ennui, Harrison composed this song on a Hammond organ that he found in their rented house on Blue Jay Way.

The recorded track itself uses backward tapes and has a heavily "flanged" effect due to the use of ADT (artificial double tracking), which uses tape delay to make a delayed copy of an audio signal that is then played back at slightly varying speed and combined with the original. The technique was and invented by Abbey Road engineers in 1966, designed to enhance the sound of voices or instruments during the mixing process.

"Your Mother Should Know"	Recorded: August 22/23, September 16/29, 1967	McCartney *(vocals, backing vocals, bass guitar, piano)*	Lennon *(backing vocals, organ)*
(Lennon-McCartney)	US Release: November 27, 1967	Harrison *(backing vocals, guitar)*	Starr *(drums, tambourine)*

Harking back to the days of music hall, this McCartney track takes its title from the screenplay of the 1961 British New Wave drama movie, *A Taste of Honey*, itself covered by The Beatles on their debut album. Recorded at Chappell Recording Studios in London, it was the last time Brian Epstein attended a Beatles recording session before his untimely death in August 1967. Interestingly, the band tried another take of the song back at Abbey Road studios, complete with harmonium and military-style snare drum pattern, which can be heard on *Anthology 2*. But, for whatever reason, this remake was completely abandoned and McCartney reverted back to the Chappell Studios version.

Magical Mystery Tour (1967)

"I Am The Walrus"	Recorded: September 5/6/27/28/29, 1967	McCartney *(backing vocals, bass guitar)*	Lennon *(double-tracked vocals, electric piano)*
(Lennon-McCartney)	US Release: November 27, 1967	Harrison *(backing vocals, lead guitar)*	Starr *(drums)*

Additional contributors: Various unnamed musicians

As an indication of the band's continuing status in the UK, the closing ceremony of the London 2012 Summer Olympics was held on August 12 in the Olympic Stadium, London. The event was dubbed "A Symphony of British Music." Up front and center at the ceremony's start was the face of John Lennon, which appeared on the big screens and was accompanied by the Liverpool Philharmonic Youth Choir and the Liverpool Signing Choir as they sang "Imagine" and a bust of Lennon's face was created.

A little later in the performance, a psychedelic bus entered the stadium with "colorful" English raconteur, Russell Brand, sitting on top singing "I Am The Walrus."

Legendary guitarist of Manchester band The Smiths, Johnny Marr, described "I Am The Walrus" as his favorite Beatles song in *Mojo* magazine of July 2006:

> As a record it stands outside of pretty much everything. It's beyond its form, really. It goes beyond what people think of as The Beatles and the '60s, but it also goes beyond what people think of as pop music as entertainment.... The first time I really discovered it I was fourteen.... It was compelling and terrifying at the same time, for a youngster, and there was this feeling of an entirely mysterious other world but with a weirdly strange familiarity to it at the same time.... It is genuinely an art object, aural art, and I don't think it's really overstating it to say that eventually it could be looked on as an artifact as important as a Picasso.... I like the way it's put together. It sounds like Hieronymus Bosch set to music. It's a certain kind of psychedelia that could only

have come out of the UK. It's neurotic and claustrophobic where American psych is vast, open and cinematic.

The song's origins are a fascinating lesson in British history. In the late 1960s, to some extent, Lennon shared Harrison's regard for the spiritual issues raised by both LSD use and the Hindu religion he shared. But Lennon's perspective was always tempered with a healthy dose of British cynicism. And in 1967's so-called "Summer of Love," Lennon's acerbic wit found much to feed off. At the time, the British bourgeoisie were worried about the burgeoning growth of the counterculture movement in the UK. They were only too well aware of the civil disruption and disobedience stemming from its sister movement in the US. And so, as the British Establishment always has, they tried to make examples of the movement's leading figures. The radical underground newspaper *International Times* was raided for "subversive material" by the police, and though the "MBE-inoculated" Fab Four were relatively immune, their more decadent brothers, The Rolling Stones, were fair game. Keith Richards, Mick Jagger, and Brian Jones were all arrested on drugs charges.

With the trials of The Stones imminent, Lennon and McCartney, themselves having recently confessed to taking LSD, showed their support by contributing backing vocals on the Stones' protest record *We Love You*. Sentenced to one year and three months respectively, Richards and Jagger were only released after *The Times* editor defended them in an editorial entitled "Who Breaks a Butterfly on a Wheel?" a quotation taken from Alexander Pope's poem of 1735, "Epistle to Dr Arbuthnot." Unfazed by this defeat and keen to show how petty they could be, the British establishment soon banned the massively popular and completely innocuous pirate radio stations, like Radio Caroline, despite huge opposition from the country's youth.

And so, in this heightened political context, Lennon was composing on his piano at home when he heard the distal two-note drone of a cop-car in the neighborhood. Later interpreted as musically

Magical Mystery Tour (1967)

representing mean-spirited authority, Lennon nonetheless immediately latched onto this siren seesaw as a relentless structure on which to hang his musical cascade, in the view of MacDonald, "an M. C. Escher staircase of all the natural major chords—the most unorthodox and tonally ambiguous sequence he ever devised."

Johnny Marr's comparison with Picasso is interesting. Lennon's track is something of a musical collage, using a number of discrete and disparate elements. Built around a basic rhythm track which itself has been processed, "I Am The Walrus" has refined parts for orchestra, chorus, radio program, and sound effects.

The track begins like other great recordings by the band; the intro is staggered, starting with part of the backing track, then the strings, and full percussion appearing last. George Martin's use of orchestration is inspired, bringing colored and textural highlights to underscore the drama in the song. Together with the use of chorus, the orchestral underlay brings out Lennon's imagery with musically artistic flourishes, perhaps inspiring Marr's near-perfect comparison to the Dutch painter Hieronymus Bosch. It is arguably the most surreal and singularly idiosyncratic protest song ever written. For example, the laughter which underscores "Expert-texpert choking smokers/Don't you think the joker laughs at you?" the stumbling triplets after "see how they run," the porcine score beneath "See how they smile/Like pigs in a sty/See how they snied [sic]," and the glissandi behind "I'm crying."

Indeed glissandi, the rapidly executed series of notes on the harp or piano, have a repeated role, popping up at different places and at different speeds. The use of a pre-recorded radio broadcast (Lennon's radio scan breaking into a BBC radio broadcast of Shakespeare's *King Lear*) fits remarkably well, given its relatively random nature. The judicious use of the radio, held back until halfway through the track, and its tactical appearance in the refrain that follows, perfectly exemplifies the "less is more" aesthetic. The sublime chorus is provided by The Mike Sammes Singers, a sixteen-voice choir of professional vocalists.

They are the Boschian voices, variously singing shrilly and whoopingly, "ho-ho-ho, hee-hee-hee, ha-ha-ha," "everybody's got one!" and the wonderful, "oompah, oompah, stick it up your jumper!"

And yet, on top of all of this is Lennon's voice and lyrics. The vocals are snarled so beautifully into the mic that the mix barely masks the resulting peak-distortions. Given this deliberately heavily processed delivery, which is dry, intentionally peak-distorted, and probably filtered for a narrower frequency bandwidth, it's little wonder that Johnny Marr declared, "In spite of the intensity of the production and the music, it's Lennon himself that's the most jarring; he commands this focus with just his words and . . . vibe. In the end it's a feeling that you get from 'I Am The Walrus'—something uncanny. To get the drug experience across in a little disc spinning round at 45 rpm is pretty phenomenal. And quite naughty. Given the circumstances, and given who he was and everything, Lennon deserves 100 million rock 'n' roll points just for that."

Seeping slowly through his consciousness over a few LSD-laced weekends, the lyrics went through several writing phases. According to Pete Shotton, English businessman and long-time friend of Lennon's, the original inspiration for "I Am The Walrus" was a letter from a lad at their old school, which detailed the way in which the English teachers were now analyzing Lennon's lyrics. Lennon found this hilarious, as previously the teachers at Quarry Bank, and *especially* the English teachers, had terminally dismissed him as a talentless disrupter—a snub that left deep scars. And so, Lennon and Shotton reminisced the quintessentially British nonsense chants they used to sing in the playground, "Yellow matter custard, green slop pie / All mixed together with a dead dog's eye," which of course became "Yellow matter custard dripping from a dead dog's eye," in "I Am The Walrus" along with a cascade of other Boschian phrases. But, as the lyric progressed, this string of the meaningless images grew more pointed, evolving from schoolboy nonsense into Lennon's simmering antipathy to the British establishment *per se*.

Thus, "I Am The Walrus" became an acerbic sequel to the darkly surrealistic "Strawberry Fields Forever," and "Walrus" became his definitive iconoclastic litany, a revolutionary rant which takes aim at the pillars of bourgeois sensibilities, berating education and culture, law and order, class and religion. Lennon exacts lyrical revenge on his "expert textpert" schoolmasters which then develops into a surreal assault on myopic parish-pump society—pretty little policemen in a row, like pigs in a sty, corporation tee-shirts, and the "wickedness" of merely letting "your knickers down." (With an irony Lennon must have found delicious, this parody on sexual repression was banned by the BBC for its use of the word "knickers"). Lennon even takes a sideswipe at satirizing the mid-sixties fashion for cultivating fanciful and pretentious Bob Dylan-induced psychedelic lyrics. The tirade climaxes as the doyens of bourgeois morality are found beating up Lennon's fellow surrealist rebel ("Man, you should have seen them kicking Edgar Allan Poe"). In the face of this onslaught, the only trace of our peaceable and mostly inclusive author is to be found in the song's opening line, "I am he / As you are he / As you are me / And we are all together."

"Hello, Goodbye"	Recorded: October 2/19/25; November 2, 1967	McCartney *(double-tracked vocals, bass guitar, piano, bongos, conga)*	Lennon *(backing vocals, lead guitar, organ)*
(Lennon-McCartney)	US Release: November 27, 1967	Harrison *(backing vocals, lead guitar)*	Starr *(drums, maracas, tambourine)*

Additional contributors: Various viola players

During the drug-influenced Summer of Love, dreamy, childlike lyrics were all the rage. Syd Barrett's Pink Floyd comes to mind, whose debut album, *Piper at the Gates of Dawn*, was being recorded at the same time, in the same studio, as *Sgt. Pepper*. McCartney's catchy "Hello, Goodbye" started life as a random exercise with alternate notes struck on a harmonium while playing a word-association game, with McCartney using his usual melodic agency over the result, and paying particular care to the bassline which, by this stage, was always mixed up high and recorded on its own track. (In Barry Miles' *Many Years From Now*, McCartney tries to put a deeper spin on the track, saying, "It's just a song of duality, with me advocating the more positive. 'You say goodbye, I say hello. You say stop, I say go.' I was advocating the more positive side of the duality, and I still do to this day.")

Nonetheless, the song was another huge success, spending seven weeks at the top of the UK chart, the band's longest residency at number one since 1963's "She Loves You." However, this success may also be seen as symptomatic of the growing divide between pop and its burgeoning brother in hard rock. Over the next few years, promising singer-songwriters with the new experimental spirit would move en masse into the album market as the singles chart started to decline.

"Strawberry Fields Forever"	Recorded: November 24/28–29; December 8-9/15/21–22, 1966	McCartney *(bass guitar, bongos, Mellotron)*	Lennon *(vocals, guitars, piano)*
(Lennon-McCartney)	US Release: February 13, 1967 (single)	Harrison *(guitar, svarmandal, timpani)*	Starr *(drums, maracas)*

Additional contributors: Various unnamed musicians

After The Beatles' last ever concert, at San Francisco's Candlestick Park, Brian Epstein promised that there would be no more touring. With open-ended recording sessions and no budget limit in sight, the band were now even more liberated in their work. As we have seen, they began the recording sessions for *Sgt. Pepper* with a focus on a semi-autobiographical series of songs which aimed to develop the resonances of haunting and nostalgic records like "In My Life" and "Eleanor Rigby." "Strawberry Fields Forever" was the first composition of this new venture. In the guise of an orphan at Strawberry Field, a girls' reform school near Lennon's childhood home in Woolton, Liverpool, the character in "Strawberry Fields" explores feelings too

Magical Mystery Tour (1967)

intense or personal to express. The song actually has two expressions: chiming with Lennon's personal journey, it's firstly a study in indeterminate identity, laced with the loneliness of the rebel against the adult world. And it's also a wistful yearning for a carefree childhood of exploration and play: the visionary strawberry fields of Lennon's mind.

To capture this haunting song about childhood memory, fifty-five hours of studio time were spent perfecting the track. The creation of new techniques was painstakingly developed from scratch by George Martin and the Abbey Road engineers. For example, Lennon told Martin he wanted to combine the "dreamy opening mood" of one take of "Strawberry Fields" with the "energetic groove" of another; the two takes having been recorded a week and a half apart. However, the two performances are not only made up of different instruments playing at different volumes, they're also in different keys and, most challenging of all, the takes were performed at different speeds. Nowadays, solving such an issue could be done on a phone app. But in 1966, it had never even been tried. George Martin wondered, *What if the faster "energetic groove" could be slowed down so it matched the "dreamy opening" take in terms of speed and musical key?* And so, they simply *invented* a device which manipulated the supply of electricity running the magnetic tape track of the "energetic groove" by a regulated amount. In the track's final version, you can't even hear the join between the two edited pieces.

"Strawberry Fields Forever" was the first recording that unveiled the extraordinary soundscapes the band had been evolving during the *Sgt. Pepper* sessions.

"It has this ghostly quality, and the slowed-down Lennon voice, which really dramatized the idea of 'maturing,'" American multi-instrumentalist musician Mark Oliver Everett said in *Mojo* magazine, "There's also a scariness that offsets the nostalgic childhood thing, and I think that's partly why it stands the test of time. From a musician's perspective, it's stunning, a spine-tingler. . . . It really raises the hair on your arms. I love the conversational tone of the lyrics—'that is,' 'I think,' 'you know,' 'ah yes'—that's unusual for a pop song, even now. It's one of those John Lennon songs that has a preordained quality, as if it just came from the sky."

"Penny Lane"	Recorded: December 29/30, 1966; January 4-6, 9-10, 12, 17, 1967	McCartney *(vocals, bass guitar, pianos, effects, harmonium, tambourine)*	Lennon *(backing vocals, guitar, pianos, congas, claps)*
(Lennon-McCartney)	US Release: February 13, 1967 (single)	Harrison *(backing vocals, guitar)*	Starr *(drums, handbell)*

Additional contributors: Various unnamed musicians

In *The Lyrics*, McCartney sites Dylan Thomas's "Under Milk Wood" as a big influence on the writing of "Penny Lane." Thomas's radio drama is a portrait of a Welsh town through a cast of characters, first played in 1953, but as McCartney says, "It was [still] very much in the air . . . there had been a new radio version of it in 1963 and a television version in 1964."

Like "Under Milk Wood," "Penny Lane" is also a docudrama in sound. McCartney describes in *The Lyrics* how the "barber [is] showing photographs" where "all the members of The Beatles had [a] hair cut there at one time or another"; "the bank, the fire station, the church I used to sing in, and [the place] where the girl stood with the tray of poppies as I waited for the bus. That pretty nurse. I remember her vividly . . . she had a tray full of paper poppies . . . and 'though she feels as if she's in a play / She is anyway,'" which McCartney also describes as "very sixties . . . a commentary on its own method."

Dylan Thomas's characters in "Under Milk Wood" bear resemblance to those in "Penny Lane," in that they are far from normal and "a bit wonky."

"There's something a bit strange about them," McCartney says in *The Lyrics*.

Magical Mystery Tour (1967)

Musically, *Penny Lane*'s blue-sky bonhomie is one of the very best examples of McCartney's vertically flexible melody and harmony. The track's lush production uses brass and piccolo, as well as flutes, a bass fiddle, and a fire bell, with voices mimicking sirens, and barely audible oboes. The track's musical highlight is the piccolo trumpet solo, which McCartney was inspired to include after seeing trumpeter David Mason play the instrument during a BBC TV broadcast of Bach's second Brandenburg Concerto. At first, Dave Mason tried to convince him that the suggested solo was out of the range of the piccolo trumpet, McCartney explains to Rick Rubin in *McCartney, 3, 2, 1*.

"And I kinda give him a look like, 'Yeah, you can do it,'" he recalled with a smirk. "So he plays it, and it haunted him for the rest of his life!"

George Martin said "the result was unique, something which had never been done in rock music before," while biographer Jonathan Gould describes the result as "impossibly high and bright," suggesting the solo is a kind of "neo-Baroque pastiche of every fanfare ever blown."

To help launch their new single, the double A-sided "Penny Lane" and "Strawberry Fields Forever," the band made a promotional film which was one of the first examples of what we now know as the music video. The film for "Penny Lane" includes footage of Liverpool, the band members dressed in matching red tunics, as well as "the shelter on the roundabout," and a fireman riding a white horse.

"Baby, You're A Rich Man"	Recorded: May 11, 1967	McCartney *(harmony vocals, bass guitar, piano)*	Lennon *(double-tracked vocals, harmony vocal, clavioline, piano)*
(Lennon-McCartney)	US Release: July 17, 1967 (single)	Harrison *(harmony vocals, guitar, claps)*	Starr *(drums, maracas, tambourine, claps)*

Recorded in just six hours at Olympic Sound Studios, "Baby, You're A Rich Man" was originally penciled in for inclusion on the animated *Yellow Submarine* movie soundtrack. Even though it does not appear on the soundtrack album, it *does* appear in the film. It was also released as the B-side to the band's single "All You Need Is Love." The song was spliced together from two separate ideas by Lennon and McCartney. Lennon's falsetto verse was originally entitled "One of the Beautiful People," while McCartney's monotone chorus provided the "baby, you're a rich man" refrain.

Lyrically, the "beautiful people" could be a reference to the 1960s hippie culture, but is more likely a gentle satire on the social standing of the band's manager, Brian Epstein. The sentiment of non-material wealth expressed in the song is a topic that Lennon would revisit four years later with "Imagine."

Musically, the sound is dominated by a clavioline on its oboe setting, which is played by Lennon. This is a three-octave monophonic keyboard that lends the track its Indian flavor. The Olympic Sound Studios sound engineer, Eddie Kramer, who worked on the track noted that "the energy level was so intense . . . that you were riding wave upon wave of amazing creativity. It was like watching a well-oiled machine. Just incredible."

McCartney recalls in *Many Years From Now* that Keith Grant mixed it "instantly, right there."

"He stood up at the console as he mixed it, so it was a very exciting mix, we were really quite buzzed."

Interesting note: Mick Jagger was at the session, and it is possible he contributed backing vocals to the chorus fade-out.

"All You Need Is Love"	Recorded: June 14/19/23–25, 1967	McCartney *(harmony vocals, bass guitar, string bass)*	Lennon *(vocals, banjo, harpsichord)*
(Lennon-McCartney)	US Release: July 17, 1967 (single)	Harrison *(harmony vocals, guitar, violin)*	Starr *(drums)*

Additional contributors: Various unnamed musicians

The anthemic "All You Need Is Love" was composed by Lennon for *Our World,* a live television broadcast that networked two dozen countries by global satellite on June 25, the same month as the release of *Sgt. Pepper*.

A real flower-power "happening," the live performance featured The Beatles, sat on high-stools, wearing huge old-school earphones, as if to underline this is going out live across the world, while their string accompaniment awaits, also wearing clunky cans, while friends of the band, including Mick Jagger and Keith Richards of The Rolling Stones, Keith Moon of The Who, Eric Clapton, Marianne Faithfull, and Graham Nash of The Hollies, were all sat on the floor in colorful "costumes."

The song's backing track had been prepared in advance and begins with the first few bars of the French national anthem and later in the coda includes Glenn Miller's "In The Mood," "Greensleeves," "Invention No. 8 in F major" by Bach, "Prince of Denmark's March," and the band's own songs "Yesterday" and "She Loves You," whose inclusion Lennon had improvised during rehearsals.

Some musicologists, including Alan Pollack, consider the use of the "She Loves You" refrain as some kind of signal from The Beatles, which served a similar purpose to the wax models of the band depicted on the cover of *Sgt. Pepper*, as if the band were deliberately distancing themselves from their past.

The Beatles (White Album) (1968)
Postmodern Perfection

"The Beatles' naïve and aggressively experimental musicianship propelled their most fractured and divisive project into a kind of accidental perfection. Fifty years later, the record is still good, still indelible, still as clean and pure as its sleeve, requiring no explanation or description beyond the band's name. As the first century of electricity and world war recedes . . . these ninety-four minutes endure, preserving the instant that rock joined the pantheon of the highest arts."

—Jordan Orlando, "The Accidental Perfection of the Beatles' White Album," *The New Yorker* (2018)

The Beatles (White Album)	Released: November 22, 1968	Recorded: May 30–October 14, 1968	Duration: 93:33
Producer: George Martin	Studio: EMI and Trident, London	Label: Apple	Tracks: 11

Track Listing

Side One

No.	Title	Lead Vocals	Length
1	"Back In The U.S.S.R."	McCartney	2:43
2	"Dear Prudence"	Lennon	3:56
3	"Glass Onion"	Lennon	2:18
4	"Ob-La-Di, Ob-La-Da"	McCartney	3:08
5	"Wild Honey Pie"	McCartney	0:52
6	"The Continuing Story Of Bungalow Bill"	Lennon	3:14
7	"While My Guitar Gently Weeps"	Harrison	4:45
8	"Happiness Is A Warm Gun"	Lennon	2:47

(Continued)

The Beatles (White Album) (1968)

Side Two			
9	"Martha My Dear"	McCartney	2:28
10	"I'm So Tired"	Lennon	2:03
11	"Blackbird"	McCartney	2:18
12	"Piggies"	Harrison	2:04
13	"Rocky Raccoon"	McCartney	3:33
14	"Don't Pass Me By"	Starr	3:51
15	"Why Don't We Do It In The Road?"	McCartney	1:41
16	"I Will"	McCartney	1:46
17	"Julia"	Lennon	2:57

All songs written by Lennon-McCartney except tracks 7 and 12, written by Harrison, and track 14, written by Starkey

Track Listing			
Side Three			
No.	Title	Lead Vocals	Length
1	"Birthday"	McCartney, with Lennon	2:42
2	"Yer Blues"	Lennon	4:01
3	"Mother Nature's Son"	McCartney	2:48
4	"Everybody's Got Something To Hide Except Me And My Monkey"	Lennon	2:24
5	"Sexy Sadie"	Lennon	3:15
6	"Helter Skelter"	McCartney	4:30
7	"Long, Long, Long"	Harrison	3:08
Side Four			
8	"Revolution 1"	Lennon	4:15
9	"Honey Pie"	McCartney	2:41
10	"Savoy Truffle"	Harrison	2:54
11	"Cry Baby Cry"	Lennon, with McCartney	3:02
12	"Revolution 9"	Various	8:22
13	"Good Night"	Starr	3:14

All songs written by Lennon-McCartney except tracks 7 and 10, written by Harrison

Postmodern Perfection

Partly in an attempt to strip away the "excess" of *Sgt. Pepper*, the band sought a sojourn in India, under Maharishi Mahesh Yogi, guru to stars and celebrities, who would later be exposed by Lennon as "Sexy Sadie."

Harrison, in particular, was keen to get back to musical basics and strip away not just the psychedelic symbolism of *Sgt. Pepper* but also remove other layers of Beatles mythology. Even if the Maharishi's lectures turned out to be a letdown, the locale was so inspiring that Lennon wrote to Ringo, who had already tired of the tedium and returned to London, saying, "We've got about two LPs worth of songs now, so get your drums out."

And so, the *White Album* was born. Thirty varied and variable tracks. Picturesque topics were replaced by the dystopian "Piggies." Romancing the meter maid was dumped for drug-induced imagery. And the world of surrealist landscapes gave way to visions of an entropic world growing in chaos. As McCartney says on the sleeve notes of a later edition of *The Beatles*: "We had left Sgt. Pepper's band to play in his sunny Elysian Fields and were now striding out in new directions without a map."

The *White Album* is a sprawling set of songs—a kind of concept album which echoes the band's indifference to corporate concerns after Brian Epstein's passing. The music hangs together as an album simply because of the sequencing skills of Lennon, McCartney, and George Martin, who conjured up the running order in a continuous twenty-four-hour effort between October 16 and 17, 1968, the band's longest single session.

"By packaging thirty new songs in a plain white jacket, so sparsely decorated as to suggest censorship," Richard Goldstein wrote in his *New York Times* review, "The Beatles ask us to drop our preconceptions about their 'evolution' and to hark back."

The white album has since become known for its kaleidoscopic style and diverse range of genres. The set includes blues, folk, ska, country rock, music hall, early metal, and the avant-garde, viewed

The Beatles (White Album) (1968)

by critics as a postmodern work and one of the greatest albums of all time. As *The Sunday Times* music critic wrote, "Musically, there is beauty, horror, surprise, chaos, order; and that is the world, and that is what the Beatles are on about: created by, creating for, their age."

Track by Track: *The Beatles (White Album)*

"Back In The U.S.S.R."	Recorded: August 22/23, 1968	McCartney *(double-tracked vocals, backing vocals, piano, lead guitar, drums, percussion, claps)*	Lennon *(backing vocals, lead guitar, bass guitar, percussion, claps)*
(Lennon-McCartney)	UK Release: November 22, 1968	Harrison *(backing vocals, lead guitar, percussion, claps)*	

When The Beach Boys released their classic album *Pet Sounds* in 1966, it became clear that The Beatles finally had some serious competition, albeit brief. The Beach Boys had, of course, started life as a surf band, but they evolved their own musical style, partly based on the doo-wop tradition, which focused on vocal harmonies.

As McCartney writes in *The Lyrics*, "They were nicking from us. Everybody was nicking from everybody else. There was a circularity to the whole enterprise."

To "nick" back, McCartney wrote "Back In The U.S.S.R.," a song about a Russian protagonist expressing relief upon returning home to the Soviet Union after visiting the US ("Flew in from Miami Beach B.O.A.C.")

As McCartney explains about his composition, the Russian "has certainly been influenced by The Beach Boys, and Chuck Berry's 'Back in the USA' is in there too. . . . He's on a plane from Miami, after all, where he's been listening to 'California Girls' in particular, which is why our bridge section refers to how the 'Ukraine girls really knock me out.' There's a pretty blatant parody of a Beach Boys chorus in the background."

"Back In The U.S.S.R." also features a knowing reference to Ray Charles' "Georgia On My Mind," except it's the homeland of Joseph Stalin rather than Jimmy Carter or Martin Luther King Jr., and the very idea in the West that someone would sing about preferring the USSR to the USA, with lines like "show me round your snow-peaked mountains way down south," and "come and keep your comrade warm" only add to the song's ludic nature.

And yet, the parody isn't all one-sided. Though our Russian narrator sings about how lucky he is to live in the USSR, he also refers to phone-tapping in the line, "Leave it 'til tomorrow to unpack my case / Honey, disconnect the phone." Arguably, (and McCartney has argued it!) the line "Take me to your daddy's farm" is a reference to the forced collectivization of the USSR's agricultural sector between 1928 and 1940 during the ascension of Stalin.

Given The Beatles were banned in the USSR at the time, the song had the paradoxical effect of making them very popular there. And so, when McCartney eventually played "Back In The U.S.S.R." to a massive audience in Red Square in 2003, it was a moment to savor.

"Dear Prudence"	Recorded: August 28/29/30, 1968	McCartney *(backing vocals, bass guitar, piano, drums, flügelhorn, tambourine, claps)*	Lennon *(double-tracked vocals, backing vocals, guitar)*
(Lennon-McCartney)	UK Release: November 22, 1968	Harrison *(backing vocals, lead guitar, claps)*	

Prudence Farrow, sister of actress Mia Farrow, had some bad experiences with LSD in the mid-1960s and came to the Maharishi Mahesh Yogi in Rishikésh to seek enlightenment through the power of Transcendental Meditation. The Maharishi had assigned two people to keep an eye on her: George Harrison and John Lennon. Having become so obsessed with meditation and its restorative abilities, Prudence would lock herself into her room away from the rest of the attendees. Lennon and Harrison tried to coax her out of her seclusion,

The Beatles (White Album) (1968)

and the experience inspired Lennon to write a song about her, "Dear Prudence."

One of a handful of songs on the *White Album* to use the finger-picking style taught to the band by Scottish singer-songwriter Donovan, the simple, innocent lyrics and cyclical, descending chord progression are almost like a nursery rhyme. The lyrics reflected their natural, simple environment with references to "sunny skies," "clouds," and "daisy chains." The recording of the song took place without Ringo who had temporarily left the group after an argument with McCartney about the drum part on "Back In The U.S.S.R." Recorded on the luxury of 8-track, the band were able to develop their ideas and layer the arrangement. McCartney, a very capable drummer, contributed a mesmeric continuous drum fill towards the end of the song, while Harrison added an ascending counter melody on guitar. In 2015, Prudence herself was asked by *Rolling Stone* magazine what she thought of the song.

"It epitomized what the '60s were about in many ways," she said. "What it's saying is very beautiful; it's very positive. I think it's an important song."

"Glass Onion"	Recorded: September 11/12/13, 1968	McCartney *(bass guitar, piano, recorder)*	Lennon *(double-tracked vocals, acoustic guitar)*
(Lennon-McCartney)	UK Release: November 22, 1968	Harrison *(lead guitar)*	Starr *(drums, tambourine)*

Additional contributors: Various unnamed musicians

Self-reference is the idea of involving a reference to oneself, or one's actions or "works." And self-reference in art occurs when an artist refers to their work in the context of the work itself. In art, the idea is closely related to "breaking the fourth wall," the fourth wall being a performance convention that an imaginary and invisible wall divides artist from audience.

Lennon's "Glass Onion," a kaleidoscopic journey through the band's back pages, is one such self-referential artwork. Firstly, it contains references to other Beatles songs ("Strawberry Fields Forever," "I Am The Walrus," "Lady Madonna," "The Fool On The Hill," and "Fixing A Hole.") Secondly, and to underscore Lennon's choice of the song title "Glass Onion" as implying layers of transparent meaning, "Lady Madonna" also references "I Am The Walrus" ("see how they run"), which in turn references "Lucy In The Sky With Diamonds" ("See how they fly like Lucy in the Sky, see how they run").

Lennon's composition also mentions "the Cast Iron Shore," the wretched waterfront on the north side of the Mersey where rubbish from the local sewers dredges up. Given "Glass Onion" is usually interpreted as Lennon's riposte to those critics ("pseuds" as Lennon called them) who looked for hidden meanings in the band's music, and is already deliberately filled as it is with red herrings, surreal imagery ("bent backed tulips" and "dove-tail joints"), and references to past works, one can only assume the inclusion of "Cast Iron Shore" is not meant as flattery!

"Ob-La-Di, Ob-La-Da"	Recorded: July 3/4/5/8/9/11/15, 1968	McCartney *(vocals, bass guitar, vocal percussion, claps)*	Lennon *(backing vocals, piano, vocal percussion, claps)*
(Lennon-McCartney)	UK Release: November 22, 1968	Harrison *(backing vocals, acoustic guitar, vocal percussion, claps)*	Starr *(drums, bongos, vocal percussion, claps)*

Additional contributors: Three sax players

According to scientists, "Ob-La-Di, Ob-La-Da" is a near perfect pop song. And yet, it was recorded after a laborious forty-two hours in the studio with the perfectionism of its main composer (McCartney) provoking some painful creative pangs along the way, with Lennon himself dubbing his partner's tune as "Paul's granny shit."

What makes the perfect pop song? After handpicking seven hundred hit songs and analyzing the eighty thousand chord progressions contained therein, researchers at the Max Planck Institute in Germany identified some of popular music's perfect moments. The middle

The Beatles (White Album) (1968)

eight of Lou Reed's aptly titled "A Perfect Day," for example. Or the key change in The Beach Boys "God Only Knows." Or the jazzy crescendo of Nina Simone's "Sinnerman." And yet, say these scientists, one has to depend upon The Beatles to find the perfect pop song from intro to outro. The researchers provided evidence as to why the band were able to have so many successful hits, including 2023's "Now And Then," while still progressively pushing musical boundaries.

The main ingredient of a perfect recipe for a popular song is "catchiness," that elusive quality of something that's easy to recall, or simply hard to forget. And, according to neuroscientists, the key element of catchiness is surprise. It's the challenge of keeping the listener on their toes in the most appealing manner which triggers the brain's reward pathways. And so, the sweet note that shocks you is a prettily-wrapped aural surprise. Consequently, when arranged in an agreeable manner, a song composed of a seamless sequence of surprising chords is the manna from heaven that keeps us hooked, "granny shit" or not.

According to McCartney in *The Lyrics*, the song's title was inspired by, "Jimmy Scott, the Nigerian conga player whom [he] liked a lot."

"Jimmy had a couple of catchphrases he used all the time," McCartney said. "One of which was 'Ob-La-Di, Ob-La-Da, life goes on, bra.' Some people think "Ob-La-Di, Ob-La-Da" is a Yoruba phrase that means something like *'comme ci, comme ça.'* Some people think it's a phrase Jimmy Scott made up. And there are others who think 'bra' refers to a brassiere, rather than an African version of 'bro.'"

Whatever the meaning, and as the researchers at the Max Planck Institute later confirmed, "Ob-La-Di, Ob-La-Da" is the most spontaneous-sounding track on the *White Album*. There was an original version, laden with heavy acoustic guitars and some session overdubs, which McCartney had been struggling with for a few days until he gave up the ghost and decided to dump it.

Luckily, however, a stoned Lennon transformed the track by pounding out a parody music-hall piano intro at faster tempo, thereby conjuring up the Jamaican ska tribute so adored by the scientists. This became the final version—complete with its gleefully silly lyric and its punchy come-as-you-are production values—which serves to underline the researchers' point about the random things thrown into the track's mixing which reminds you that life does indeed "go on" and that the song makes you feel very much alive.

All this made "Ob-La-Di, Ob-La-Da" the most commercial track on *The Beatles*. When the other band members, somewhat fed up and pained by all the birthing effort, vetoed it as the band's next single, it was promptly cashed in by Glasgow-based band, Marmalade, who took it straight to the top of the singles chart.

"Wild Honey Pie"	Recorded: August 20, 1968	McCartney *(vocals, acoustic guitars, drums)*
(Lennon-McCartney)	UK Release: November 22, 1968	

A fragment of a sing-along, also written in Rishikésh, "Wild Honey Pie" was recorded by McCartney alone. It was recorded at the end of the "Mother Nature's Son" session and used the same corridor drum setup. McCartney overdubbed numerous guitars, vocals, and a thumped bass drum onto this spontaneous song captured in one take; the song finishes after fifty-two seconds. According to McCartney, it was only included on the album because Pattie Harrison liked it.

"The Continuing Story Of Bungalow Bill"	Recorded: October 8, 1968	McCartney *(backing vocals, bass guitar)*	Lennon *(vocals, acoustic guitar, organ)*
(Lennon-McCartney)	UK Release: November 22, 1968	Harrison *(backing vocals, acoustic guitar)*	Starr *(backing vocals, drums, tambourine)*

Additional contributors: Yoko Ono (vocals, backing vocals)

John Lennon often used scorn and laughter as raw ingredients in his artistic material. Whereas McCartney's songs were marked

The Beatles (White Album) (1968)

by bright optimism, Lennon's were often darkly acidic. So, given Lennon's predilection, it's hardly surprising that when Lennon encountered a tiger-hunting "spiritualist" in India, he should write a song about it.

"The Continuing Story Of Bungalow Bill" was written in 1968 in Rishikésh and was inspired by a young American named Richard A. Cooke III. According to Cooke's mother, the American socialite Nancy Cooke de Herrera, she and her son were on friendly terms with all band members save for Lennon, whom Nancy described as a distant "genius" who was contemptuous of her and her clean-cut, college-attending son.

According to Nancy's autobiography, *Beyond Gurus*, Lennon's satirical song arose after the Cookes departed for a tiger-killing spree before returning to the ashram to seek spiritual enlightenment. As Lennon put it in the 1980 *Playboy* interviews, "'Bungalow Bill' was written about a guy in Maharishi's meditation camp who took a short break to go shoot a few poor tigers, and then came back to commune with God. There used to be a character called Jungle Jim, and I combined him with Buffalo Bill. It's sort of a teenage social-comment song and a bit of a joke."

The song starts up with a flourish on a flamenco guitar, played from a standard Mellotron bank of pre-recorded phrases by studio engineer Chris Thomas, before launching into a narrative which is sung by all four Beatles, Ringo's then-wife Maureen, and Yoko Ono. The track was purposefully recorded with a sloppy spontaneity and a tone-shifting strangeness in three takes. Yoko's contribution, on the line "not when he looked so fierce," was the first and only time a female lead vocal appeared on a Beatles recording.

"While My Guitar Gently Weeps"	Recorded: July 25, August 16, September 3/5/6, 1968	McCartney *(backing vocals, bass guitar, piano, organ)*	Lennon *(lead guitar)*
(Harrison)	UK Release: November 22, 1968	Harrison *(double-tracked vocals, backing vocals, acoustic guitar, Hammond organ)*	Starr *(drums, tambourine)*

Additional contributors: Eric Clapton (lead guitar)

It was a hard life being George Harrison in The Beatles. A guitarist trying to develop his craft as a songwriter, his songs had to wrestle for space with Lennon and McCartney's songs on every album.

"I always had to do about ten of Paul and John's songs before they'd give me the break," he once complained.

He finally got his chance with the *White Album*, on which he managed to get a full five songs recorded. The most well-known of these is "While My Guitar Gently Weeps." Harrison started composing the music for the song on acoustic guitar in India, using a finger-picking style.

The lyrics he completed while back in England, as Harrison explains, "I wrote 'While My Guitar Gently Weeps' at my mother's house in Warrington. I was thinking about the Chinese *I Ching*, the Book of Changes.... The Eastern concept is that whatever happens is all meant to be, and that there's no such thing as coincidence—every little item that's going down has a purpose. I decided to write a song based on the first thing I saw upon opening any book—as it would be a relative to that moment, at that time. I picked up a book at random, opened it, saw 'gently weeps,' then laid the book down again and started the song."

Like so many Beatles recordings, the song was to change considerably from conception to completion. Harrison originally recorded it solo on acoustic guitar, before the band proceeded to record two further remakes, the first on Abbey Road's new eight-track recording equipment. Unhappy that the band were not taking the song seriously enough and not giving it the attention it deserved, he invited Eric Clapton to play the lead guitar solo on the track.

According to Harrison: "The next day I was driving into London with Eric Clapton, and I said, 'What are you doing today? Why don't you come to the studio and play on this song for me?' He said, 'Oh, no—I can't do that. Nobody's ever played on a Beatles record and the others wouldn't like it.' I said, 'Look, it's my song and I'd like you to play on it.' So he came in. I said, 'Eric's going to play on this one,' and it was good because that then made everyone act better. Paul got on the piano and played a nice intro and they all took it more seriously."

For the recording, Clapton used a cherry-red Gibson Les Paul guitar, nicknamed Lucy, that he had given to Harrison a month previously, and had it double-tracked to make it sound less bluesy and more Beatle-like.

"Happiness Is A Warm Gun"	Recorded: September 23-25, 1968	McCartney *(backing vocals, bass guitar)*	Lennon *(double-tracked vocals, backing vocals, lead guitar)*
(Lennon-McCartney)	UK Release: November 22, 1968	Harrison *(backing vocals, lead guitar)*	Starr *(drums, tambourine)*

Like the man himself, Lennon's compositions were often complex. "Happiness Is A Warm Gun" is a musical meld of four different fragments: the finger-picking folk style of the intro, the bluesy phrasing of the "I need a fix" section, the straight up rock styling of the "Mother Superior" section, and the concluding mid-fifties doo-wop outro.

All the band members loved the track, which shows in the fact that they spent fifteen hours and almost one hundred takes perfecting it. Hardly surprising when you consider the song is the most metrically irregular song they ever recorded.

The origin of the song itself is also fragmentary. The band's press officer, Derek Taylor, said that the first part was pieced together from word associations he and Lennon had conjured at random: the lizard on the windowpane (a memory of Los Angeles), the man in the crowd (a real character who stuck small mirrors onto the toe-caps of his boots so he could look up women's skirts at soccer matches), and the song's title (famously a slogan of America's National Rifle Association), which Lennon saw in a gun magazine.

And so, this dark amalgam of lyrical influences and intricate musicality was pieced together to make a song which Lennon rated one of his best and most ambiguous tracks. As Tori Amos said about the song in *Mojo* magazine in 2006, "It's hard to write a song that has a social commentary. A song that can make a statement without preaching is rare, and this one did it. The Beatles had the ability to make you think about the world, not just your own little world. They could put the microcosm and macrocosm in the same song. They sang of drugs and guns without telling me what to feel about it. That's genius."

"Martha My Dear"	Recorded: October 4/5 1968	McCartney *(double-tracked vocals, bass guitar, piano, lead guitar, drums, claps)*	
(Lennon-McCartney)	UK Release: November 22, 1968		

Additional contributors: Various unnamed musicians

"Martha My Dear" is McCartney's jazzy, blues-inflected homage to his old English sheepdog, Martha. Martha played an essential role in the development of the relationship between McCartney and Lennon. According to McCartney in *The Lyrics*, "One of the unlikely side effects was that John became very sympathetic towards me. When he came round and saw me playing with Martha, I could tell that he liked her. John was a very guarded person, which was partly where all his wit came from. He'd had a very difficult upbringing, what with his father leaving home, his uncle dying, and his mother getting killed in a traffic accident." McCartney explains that, by the time he got to know him, Lennon could be very sarcastic. It was his way of dealing with the knocks that life had dealt him. And so the two scousers were

The Beatles (White Album) (1968)

"quite into the witty put-down. But seeing me with Martha, with my guard down, all of a sudden he started warming to me. And so he let his guard down too."

A piano opens the piece, played with no pedal and closely mic'd, and runs beneath the track. In addition, we have the kind of instrumental layering which is by now a Beatles convention: light strings to underscore the piano solo, brass chords on the odd beats of each measure, with string chords on every beat, drums on the bridge, and a trumpet solo.

On the release of the *White Album*, McCartney was amused by the fact that, at the time, almost no one listening to the song knew that Martha was a dog. Having said that, McCartney also knows that, as the song proceeds, he deliberately penned the lyrics so that Martha appears to morph into a person. Nonetheless, given the song includes phrases such as "hold your head up" and "silly girl," one would certainly hope McCartney was connecting with a dog rather than a real woman.

"I'm So Tired"	Recorded: October 8, 1968	McCartney *(harmony vocals, bass guitar, electric piano)*	Lennon *(vocals, acoustic guitar, lead guitar, organ)*
(Lennon-McCartney)	UK Release: November 22, 1968	Harrison *(lead guitar)*	Starr *(drums)*

According to Lennon, three weeks of constant meditation and lectures at the camp at Rishikésh had a profoundly negative effect on him.

"'I'm So Tired' was me, in India again. I couldn't sleep, I'm meditating all day and couldn't sleep at night. The story is that. One of my favorite tracks. I just like the sound of it, and I sing it well."

It was not the first time he had visited the topic of sleep, having dwelled upon it for the song "I'm Only Sleeping" on *Revolver* two years earlier. Indeed, the two songs can be thought of as companion pieces on Lennon's state of mind during these years.

Completed in a single, lengthy sixteen-hour session, this brilliant band performance was recorded live on separate tracks, with some overdubs added later. Lyrically, "I'm So Tired" contains one of Lennon's most amusing lines: cursing English statesman and explorer Sir Walter Raleigh, the "discoverer" of tobacco, for being "such a stupid git."

At the end of the song, Lennon mutters the words "Monsieur, monsieur, monsieur, how about another one?" This was (somehow) interpreted by conspiracy theorists as "Paul is dead, man, miss him, miss him, miss him," when played backwards, and became part of the "evidence" for the infamous "Paul is Dead" myth.

"Blackbird"	Recorded: June 11, 1968	McCartney *(vocals, acoustic guitar)*	
(Lennon-McCartney)	UK Release: November 22, 1968		

"Blackbird," a solo rendition by McCartney, was written on his farm in Scotland shortly after the band's stay in Rishikésh. On the first night his future wife Linda Eastman stayed at McCartney's house in St. John's Wood, he performed the song to the fans waiting outside the gates.

In his book *Shout!: The Beatles in Their Generation*, English author Philip Norman includes an account from Beatles fan Margo Stevens, who was present at this first impromptu performance of "Blackbird":

> A few of us were there. We had the feeling something was going to happen. Paul didn't take the Mini inside the way he usually did; he parked it on the road and he and Linda walked right past us. They went inside and we stood there, watching different lights in the house go on and off. In the end, the light went on in the Mad Room, at the top of the house, where he kept all his music stuff and his toys. Paul opened the window and called out to us, "Are you still down there?" "Yes," we said. He must have been really

The Beatles (White Album) (1968)

happy that night. He sat on the windowsill with his acoustic guitar and sang "Blackbird" to us as we stood down there in the dark.

The music of "Blackbird" was based on Bach's "Bourrée in E minor," which intrigued McCartney with its "harmonic thing between the melody and the bass line."

The words of the song, however, were inspired by the civil rights movement in America. As McCartney explains in *The Lyrics*, "I am very conscious that Liverpool was a slave port, and also that it had the first Caribbean community in England. So we met a lot of Black guys, particularly in the music world. I'm thinking in particular of Lord Woodbine, a calypso singer and promoter who ran a couple of joints in Liverpool, including the New Cabaret Artists' Club, where he hosted The Silver Beetles. Then there was Derry of Derry and the Seniors, a band that had paved the way for us in Hamburg. At the time in 1968 when I was writing 'Blackbird,' I was very conscious of the terrible racial tensions in the US. The year before, 1967, had been a particularly bad year, but 1968 was even worse. The song was written only a few weeks after the assassination of Martin Luther King Jr. That imagery of the broken wings and the sunken eyes and the general longing for freedom is very much of its moment. [Soon after] I started to do a lot more in the way of contextualizing the songs in my regular concerts by telling some of the stories behind them. I think audiences really appreciate finding an angle into the planetary atmosphere of the song, by getting to see the far side of the moon."

The song turned up again fifty-five years later on Beyoncé's *Cowboy Carter* album, with her production team using the instrumental elements, McCartney's acoustic guitar and foot tapping, from the original master recording. Beyoncé's cover was in full recognition of the fact that McCartney had penned the original as a tribute to the Little Rock Nine, a group of students who had faced racial discrimination after starting at the (previously) all-white Little Rock high school in 1957. McCartney's lyrics are steeped in metaphor and symbolism. He told *GQ* magazine in 2018 how "in England, a bird is a girl [or was in 1960s slang], so I was thinking of a black girl going through this—you know, now is your time to arise, set yourself free, and take these broken wings."

And the equally oblique reference to the Little Rock students themselves is in the line "all your life, you were only waiting for this moment to be free."

"Piggies"	Recorded: September 19/20; October 10, 1968	McCartney *(bass guitar)*	Lennon *(tape loops)*
(Harrison)	UK Release: November 22, 1968	Harrison *(vocals, guitar)*	Starr *(tambourine)*

Additional contributors: Various unnamed musicians

George Orwell's 1945 novella, *Animal Farm*, is a beast fable, a satirical and allegorical tale in which a group of anthropomorphic farm animals rebel against their human farmer and take over the farm. At first all is well, but soon a dictatorship is set up under a pig called Napoleon and the pigs elevate themselves above the other animals to become a new ruling class.

George Harrison's "Piggies" is a misanthropic and satirical stab at the heart of capitalistic human society in which he depicts the ruling class as pigs. Originally written in 1966, the year he also created (the not quite as caustic) "Taxman," "Piggies" begins with the classical connotations of harpsichord and cello, underlining the fact that Harrison's target is bourgeois society. The noise of pigs snorting on the track was assembled by Lennon using tape effects from the Abbey Road collection.

As engineer Stuart Eltham is quoted as saying in Mark Lewisohn's *The Complete Beatles Recording Sessions*, "There's a tape called *Animals And Bees (volume 35)* which includes pigs. It's from an old EMI 78 rpm record and The Beatles may have used a combination of that and their own voices. That always works well—the new voices hide the 78 rpm scratchiness, the original record hides the fact that some of the sounds are man made."

The Beatles (White Album) (1968)

Meanwhile, we find that, lyrically, the "bigger piggies in their starched white shirts" are forever making sure the "little piggies" are "crawling in the dirt" while the bigger piggies cannibalistically clutch their "forks and knives to eat their bacon." Lyrically assisted by Lennon and Harrison's mother, Louise, by the time "Piggies" was unleashed upon the world in November of 1968, the American hippie counter-culture was clashing violently with their ruling class. Although clearly derived from hippie ideals, a world away from pinstriped businessmen, elements of Harrison's potent misanthropic song were adopted as a mantra of revolutionary mass-murder by the deranged Charles Manson.

Inspired by Louise Harrison's line, "What they need's a damn good whacking," Manson adopted "Piggies" as one of the tracks to justify attacks on the "White" bourgeoisie. Believing that the Beatles were instructing him through their music, Manson saw these attacks as the prelude to an apocalyptic racial war between the establishment and the Black community that would leave Manson and his disciples to rule America on counterculture principles in some kind of bizarre echo of Orwell's *Animal Farm*.

Following the infamous murder of Sharon Tate, the pregnant wife of film director Roman Polanski, Manson disciple Susan Atkins used Tate's blood to daub "Pig" on the front door of the murder scene. A day later, after the murders of Leno and Rosemary LaBianca using knives and forks, another Manson disciple, Patricia Krenwinkel, used the victims blood to write "Death to pigs" on the walls.

"Rocky Raccoon"	Recorded: August 15, 1968	McCartney *(vocals, acoustic guitar)*	Lennon *(backing vocals, bass guitar, harmonica, harmonium)*
(Lennon-McCartney)	UK Release: November 22, 1968	Harrison *(backing vocals)*	Starr *(drums)*

Additional contributors: George Martin (piano)

As irrepressible Liverpudlians, and doubtless in no small part due to their seasoned experience of playing in Hamburg, both Lennon and

McCartney were skilled at the art of jokey ad-libs in various musical styles. Witness "Rocky Raccoon," a kind of silent-movie comedy song about cowboys.

Underlining once more the Scouse and Irish liking for country-and-western music, the song was previously a private joke but in the new no-holds-barred atmosphere at Abbey Road, "Rocky Raccoon" "checked into" the room and "proceeded to lie on the table" of songs for the new album. George Martin was of the opinion that the band's apparently less fussy attitude to selecting tracks for *The Beatles* was simply down to recording fillers so they might fulfill their EMI contract faster.

Consequently, Martin tried to persuade the band to ditch half of their new material and issue one strong album rather than the "weaker" double they eventually released. And so "Rocky Raccoon," which started life as a musical jam with Lennon and Scottish singer Donovan in India, was transformed into a kind of late '60s "Ballad of Buster Scruggs" for the hippie generation. Taped in one night, with its use of acoustic guitar, supplemented at times by combinations of drum kit, bass, barrelhouse piano faked with ADT and varispeed, harmonica, harmonium, and backing vocals, the production of "Rocky Raccoon" variously adds and subtracts instruments to articulate atmosphere and provide a sense of forward progress.

"Don't Pass Me By"	Recorded: June 5/6, July 12/22, 1968	McCartney *(bass guitar, piano)*	
(Starkey)	UK Release: November 22, 1968		Starr *(vocals, drums, sleigh bell, piano)*

Additional contributors: Jack Fallon (violin)

The first Ringo composition to be recorded by the band, this country-and-western song, with touches of bluegrass, was possibly written as early as 1963. There is some debate over the exact date, but there are references to Ringo's "first song" in radio interviews of that year.

The Beatles (White Album) (1968)

According to Ringo, "I wrote 'Don't Pass Me By" when I was sitting round at home. It was great to get my first song down, one that I had written. It was a very exciting time for me, and everyone was really helpful, and recording that crazy violinist was a thrilling moment."

Recorded with McCartney on piano, in addition to his usual bass, and with Ringo on drums, the "crazy violinist" at the end was provided by session musician Jack Fallon. Fallon was not happy with the quality of his playing towards the very end of his part.

"I thought that they had had enough, so I just busked around a bit." Fallon said in *The Complete Beatles Recording Sessions*.

"When I heard it played back at the end of the session, I was hoping they'd scrub that bit out, but they didn't, so there I am on record, scraping away! I was very surprised they kept it in. It was pretty dreadful."

Fallon was also the person who had booked the band into their first professional engagement in southern England (Stroud, in fact) in March 1962, ten weeks before the group's EMI audition.

"Why Don't We Do It In The Road?"	Recorded: October 9/10, 1968	McCartney *(vocals, bass guitar, acoustic guitar, lead guitar, piano, claps)*	
(Lennon-McCartney)	UK Release: November 22, 1968		Starr *(drums, claps)*

Spontaneously recorded towards the end of the *White Album* sessions, with only McCartney and Ringo present, this bluesy number was inspired by the sight of copulating monkeys that McCartney had observed while meditating in Rishikésh.

As he explains, "I was up on the flat roof meditating and I'd seen a troupe of monkeys walking along in the jungle and a male just hopped on to the back of this female and gave her one, as they say in the vernacular. . . . There is an urge, they do it, and it's done with. And it's that simple. We have horrendous problems with it, and yet animals don't. 'Why Don't We Do It In The Road?' was a primitive statement to do with sex or to do with freedom really. I like it, it's just so outrageous that I like it."

Recorded over two days, the first saw McCartney lay down the basic track of acoustic guitar, piano, and percussion while Ringo added drums and handclaps on the second day. The raunchy and earthy feel on the track greatly appealed to Lennon, who was hurt by not being asked to contribute to the song.

"He recorded it by himself in another room," Lennon later explained, referring to McCartney. "That's how it was getting in those days. We came in and he'd made the whole record. Him drumming. Him playing the piano. Him singing. I enjoyed the track. Still, I can't speak for George, but I was always hurt when Paul would knock something off without involving us. But that's just the way it was then."

"I Will"	Recorded: September 16/17, 1968	McCartney *(vocals, acoustic guitars, "vocal bass")*	Lennon *(percussion)*
(Lennon-McCartney)	UK Release: November 22, 1968		Starr *(cymbals, bongos, maracas)*

In *The Lyrics*, McCartney compares himself to Alan-a-Dale, the minstrel wandering around Sherwood Forest in the Robin Hood legend. McCartney is specifically referring to the song "I Will," and calls himself a troubadour, who is infatuated with singing songs about "the joy of love" as the "mightiest, strongest force on the planet."

Little surprise, then, that American musicologist Alan Pollack describes "I Will" as "a pink-edged daisy of a song" which is "inscrutably hymn-like, even religious in tone." The creation of "I Will" began in February of 1968, when McCartney was in India. Attempting to "capture that feeling of first being in love," as usual he found that "the music came together quite quickly" (it still being one of the favorite melodies of McCartney's), but the words took a little longer.

Though McCartney was with actress Jane Asher at the time, he insists the song isn't "addressed to, or about, Jane. When I'm writing,

The Beatles (White Album) (1968)

it's as if I'm setting words and music to the film I'm watching in my head. It's a declaration of love, yes, but not always to someone specific."

If the lyrics to "I Will" took a little longer, the recording of the song took longer still. The soothing verse-chorus melody and susurrous internal rhymes needed an enervating sixty-seven takes due to the acoustic set-up which laid bare the slightest mistake. As usual with The Beatles, and especially McCartney, the apparent ease of the end result is the product of a huge amount of effort and concentration.

"Julia"	Recorded: October 13, 1968		Lennon *(double-tracked vocals, double-tracked acoustic guitar)*
(Lennon-McCartney)	UK Release: November 22, 1968		

Songs by rock musicians about their dead mothers are relatively rare. A tender love song to his mother, "Julia" finds Lennon finally exorcising the pain and anger that consumed him following her death and into his twenties. Killed by a car driven by an off-duty police officer, the event devastated him and helped fuel his subsequent hatred of authority and authority figures.

The song contains subtle references to Yoko Ono, particularly within the use of the phrase "ocean child"; the name Yoko means "child of the sea" in Japanese. In many ways this song can be viewed as the transfer of Lennon's devotion and obsessive love from his dead mother to his new love, Yoko.

Written in India, the song is another which uses the finger-picking method—novel at the time, but now widely in use, it was recorded in a few takes with minimal overdubs. Lyrically, the song contains imagery adapted from the poem "Sand and Foam" by Kahlil Gibran. For example, the opening line of "Julia" is very similar to Gibran's, "Half of what I say is meaningless; but I say it so that the other half may reach you." Lennon's line "when I cannot sing my heart, I can only speak my mind" mirrors Gibran's "when life does not find a singer to sing her heart she produces a philosopher to speak her mind."

"Julia" is the only solo Lennon recording in the entire Beatles catalogue.

"Birthday"	Recorded: September 18, 1968	McCartney *(vocals, piano)*	Lennon *(vocals, backing vocal, lead guitar)*
(Lennon-McCartney)	UK Release: November 22, 1968	Harrison *(bass guitar)*	Starr *(drums, tambourine)*

This song that starts the second half of the *White Album* emerged from a jam in Abbey Road. As engineer Chris Thomas explains in Lewisohn's *Complete Beatles Recording Sessions*, "Paul was the first one in, and he was playing the 'Birthday' riff. Eventually the others arrived, by which time Paul had literally written the song, right there in the studio. We had the backing track down by about 8:30 . . . and actually finished the whole song . . . in a day!"

In Barry Miles' *Many Years From Now*, McCartney says that they "Got a riff going and arranged it around this riff. We said, 'We'll go to there for a few bars, then we'll do this for a few bars.' We added some lyrics, then we got the friends who were there to join in on the chorus. So that is 50-50 John and me, made up on the spot and recorded all on the same evening."

McCartney is referring to the fact that both Pattie Harrison and Yoko Ono were among those providing backing vocals to the song, while the effects heard towards the end of the song, especially in the "I would like you to dance" section, were made by a piano mic fed through a guitar amplifier with added effects.

In *The Lyrics*, McCartney explains that "'Birthday' is one of those songs that was written to be played in a live show [and] . . . still works well for an audience." Moreover, "Two of the lines I'd focus on are 'I would like you to dance/Take a cha-cha-cha-chance.' I'm reminded that another band that was very much to the fore then was

The Beatles (White Album) (1968)

The Who. They had a very memorable moment in 'My Generation,' which involved what we used to call a stammer or a stutter on the phrase 'fade away.' But when you go 'f-f-f-f-' on live British television, that gets people's attention. I remember that moment quite vividly. And that 'impediment,' let's call it, informed the 'cha-cha-cha' in 'Birthday,' just as 'Birthday' informed the 'cha-cha-cha-cha' in David Bowie's 'Changes.' Being a songwriter is about picking up the baton and holding it for a while and then passing it on."

"Yer Blues"	Recorded: August 28/29/30, 1968	McCartney *(bass guitar)*	Lennon *(vocals, backing vocals, lead guitar)*
(Lennon-McCartney)	UK Release: November 22, 1968	Harrison *(lead guitar)*	Starr *(drums)*

In 1968, the British Blues Boom was in full swing, with bands such as Fleetwood Mac and a Clapton-led Cream leading the charge. The Beatles were never really into the blues in its raw form, preferring to draw their influences from the more nuanced R&B and soul genres. Seen from this perspective, "Yer Blues" is Lennon's satirical comment on British "white men singing the blues," complete with tongue-in-cheek two-note guitar solos and blues shuffles.

Written high up in the beautiful mountains of India, the song finds Lennon "trying to reach God and feeling suicidal." The "girl" of the song is likely to be Yoko, who was regularly sending Lennon postcards from England. The raw power of the recording was captured in a small stripped-down annex next to studio 2 in Abbey Road. Studio engineer Ken Scott remembers the session: "The room next door was tiny, where the four-track tape machines were once kept, and it had no proper studio walls or acoustic set-up of any kind . . . we literally had to set it all up—them and the instruments—in this minute room. That's how they recorded 'Yer Blues,' and it worked out great!"

The lack of soundproofing and acoustic isolation meant that vocals and guitars spilled over onto other tracks, giving the track the feel of a live performance. Ringo affectionately recalls, "'Yer Blues,' on the

White Album? You can't top it. It was the four of us. That is what I'm saying: it was really because the four of us were in a box, a room about eight by eight, with no separation. It was this group that was together; it was like grunge rock of the sixties, really—grunge blues."

"Mother Nature's Son"	Recorded: August 9/20, 1968	McCartney *(double-tracked vocals, acoustic guitars, drums, timpani)*
(Lennon-McCartney)	UK Release: November 22, 1968	

Additional contributors: Two unnamed trumpet players and two unnamed trombone players

Both Lennon and McCartney wrote songs inspired by a Maharishi Yogi lecture about nature. Lennon's was a piece called "I'm Just A Child Of Nature," which later became "Jealous Guy." McCartney's was "Mother Nature's Son."

The song is a paean of all things pastoral, and a musical close cousin of "The Fool On The Hill," possessing a similar mood and musical feeling of remoteness. The song flaunts a buoyant McCartney take on all the musical parts. George Martin's layered arrangement is subtly choreographed. McCartney's acoustic guitar is a constant on the track, with brass plussed in for much of the mid-section, and an unusually tiny contribution from percussion. McCartney paid particular attention to the position of the drums, eventually ending up recording them partway down a corridor outside the studio.

While the pop mainstream mulled over more metropolitan matters, the growing underground movement experimented musically through countercultural tropes, which included a vibrant back-to-nature conviction. The song was recorded during an often-fractious period for the band. As Ken Scott was quoted as saying in Mark Lewisohn's *The Complete Beatles Recording Sessions*, "Paul was downstairs going through the arrangement with George and the brass players. Everything was great, everyone was in great spirits. It felt really good. Suddenly, halfway through, John and Ringo walked in and you could cut the atmosphere with a knife. An instant change. It was like that for ten minutes and then as soon as they left it felt great again. It was *very* bizarre."

"Everybody's Got Something To Hide Except Me And My Monkey"	Recorded: June 26/27; July 1/23, 1968	McCartney *(backing vocals, bass guitar, percussion, claps)*	Lennon *(vocals, guitar, percussion, claps)*
(Lennon-McCartney)	UK Release: November 22, 1968	Harrison *(backing vocals, lead guitar, percussion, claps)*	Starr *(drums, percussion, claps)*

According to Lennon, the song was written about his relationship with Yoko Ono, "It was about me and Yoko. Everybody seemed to be paranoid except for us two. All this sort of madness is going on around us because we just happened to want to be together all the time."

However, McCartney believed that the "monkey" in the song's title was a reference to heroin, as it was commonly referred to by jazz musicians in the 1940s: "He was getting into harder drugs and so his songs were taking on more references to heroin. John started talking about fixes and monkeys and it was a harder terminology which the rest of us weren't into."

Musically, the song is driven along by McCartney's frantic clanging of a handbell and Harrison's exceptional, stinging guitar riffs, while Lennon's vocals are a real tour de force. Always a rock-and-roller at heart, this riff-heavy song sees Lennon returning to his roots following the psychedelia of *Revolver* and *Sgt. Pepper*. The lyrics are peppered with the favorite sayings of the Maharishi. Harrison revealed that "everybody's got something to hide" and "Come on. It's such a joy. Take it easy. Take it as it comes. Enjoy!" were common phrases from his lectures and teachings.

The performance is one of the tightest and rockiest they ever committed to tape. Interesting to note that the band experimented with a new working method on the *White Album*: They would rehearse and rehearse several times before recording the final version of the song to tape in order to develop their ideas and let their creativity flow. This was the case with "Monkey." It is also interesting to note that some of the songs on the album took over one hundred takes to perfect.

"Sexy Sadie"	Recorded: July 19/24; August 13/21, 1968	McCartney *(backing vocals, bass guitar, piano)*	Lennon *(double-tracked vocals, backing vocals, acoustic guitar, electric rhythm guitar, Hammond organ)*
(Lennon-McCartney)	UK Release: November 22, 1968	Harrison *(backing vocals, lead guitar)*	Starr *(drums, tambourine)*

While the band were at Rishikésh, a rumor started circling that the Maharishi had made sexual advances on a female member of the meditation camp. John's first wife, Cynthia Lennon, claimed that the rumor was unfounded, as there was no evidence or justification for it, but she was fueled by jealousy of the band's tech guru "Magic" Alexis Mardas.

Disillusioned with the Maharishi, John used the situation as an excuse to leave India. As he was waiting to leave, he started to write "Sexy Sadie" about the Maharishi.

"I wrote it when we had our bags packed and were leaving," Lennon said. "It was the last piece I wrote before I left India. I was just using the situation to write a song, rather calculatingly but also to express what I felt. I was leaving the Maharishi with a bad taste. You know, it seems that my partings are always not as nice as I'd like them to be."

According to Harrison, Lennon started singing the song as they drove to Delhi, "John had a song he had started to write which he was singing: 'Maharishi, what have you done?' and I said, 'You can't say that, it's ridiculous.' I came up with the title of 'Sexy Sadie' and John changed 'Maharishi' to 'Sexy Sadie.'"

The Beatles (White Album) (1968)

Legend has it that when the Maharishi asked why he was leaving, Lennon replied, "Well, if you're so cosmic, you'll know why."

"Helter Skelter"	Recorded: August 22/23, 1968	McCartney *(vocals, lead guitar, bass guitar)*	Lennon *(backing vocals, lead guitar, bass guitar, tenor sax)*
(Lennon-McCartney)	UK Release: November 22, 1968	Harrison *(backing vocals, rhythm guitar)*	Starr *(drums)*

Additional contributors: Mal Evans (trumpet)

A loud and raucous rocker, this classic is sometimes credited as the start of heavy metal and aptly named after a fairground ride. McCartney uses the up-and-down fact of the fairground ride as a metaphor for life, one moment you're feeling euphoric, the next you're feeling miserable.

According to McCartney in *The Lyrics*, the song's verses are based on "the Mock Turtle's song from *Alice in Wonderland*," including the couplet, "Will you, won't you, will you, won't you, will you join the dance?" As he further explains, "John and I both adored Lewis Carroll, and we often quoted him. Lines like 'She's coming down fast' have a sexual component, perhaps a little drug component too. A little darker." But things really got dark when Charles Manson, a year later, hijacked the song. He thought The Beatles were the Four Horsemen of the Apocalypse, and he was reading all this stuff into the lyrics. All sorts of secret meanings. Apparently, he read hell into "Helter Skelter."

The recording process for the track was something of a feat of endurance, so much so that, at the end of it, "Ringo ... shouted out something about having blisters on his fingers." The song was recorded a number of times during the sessions for the *White Album*. On one of those sessions, the July 18 session, the band recorded take three of the song, which lasted over twenty-seven minutes. This version differs from, and is generally slower than, the final album version, but may have been the origin of Ringo's blisters.

"Long Long Long"	Recorded: October 7-9, 1968	McCartney *(backing vocals, bass guitar, organ)*	
(Harrison)	UK Release: November 22, 1968	Harrison *(double-tracked vocals, acoustic guitars)*	Starr *(drums)*

Additional contributors: Chris Thomas (piano)

Described in Ian MacDonald's *Revolution in the Head* as Harrison's "touching token of exhausted, relieved reconciliation with God," this hymn-like tribute represents the spiritual heart of the *White Album*. Inspired by the chord progression of the Bob Dylan song, "Sad Eyed Lady Of The Lowlands," and the lo-fi quality of The Band's "Music from Big Pink," the rhythm track was recorded after sixty-seven takes in a marathon sixteen-and-a-half hour session.

The meditative verses are only briefly disturbed by the middle eight section with piano played by George Martin's assistant Chris Thomas. The ending of the song contains one of those magical Beatles accidents as described by Thomas: "There's a sound near the end of the song which is a bottle of Blue Nun wine rattling away on top of a Leslie speaker cabinet. It just happened. Paul hit a certain note and the bottle started vibrating. We thought it was so good that we set the mics up and did it again. The Beatles always took advantage of accidents."

To compound the effect, Harrison added a ghostly howl, while Starr provided some frenetic drumming. Chris Gerard of the international online magazine of popular culture *PopMatters* refers to the "palpable spiritual longing" conveyed in the song and describes this coda as a "weird spectral ending, with Harrison wailing like a wounded ghost while the band members rattle their instruments ominously." Quite.

The Beatles (White Album) (1968)

"Revolution 1"	Recorded: May 30-31; June 4/21, 1968	McCartney *(backing vocals, bass guitar, piano, organ)*	Lennon *(vocals, acoustic guitar, lead guitar)*
(Lennon-McCartney)	UK Release: November 22, 1968	Harrison *(backing vocals, lead guitar)*	Starr *(drums)*

Additional contributors: Various unnamed musicians

The first song to be taped for the *White Album*, Lennon's "Revolution" was recorded with the man himself lying on his back on the studio floor to get the kind of loose, breathy, and relaxed sound he was seeking.

The song was written against a backdrop of the May 1968 student uprising in Paris (which also inspired "Street Fighting Man" by The Rolling Stones), the anti-Vietnam war pitched battle outside the US embassy in Grosvenor Square, London, between police and 100,000 marchers, and just a few weeks later, the assassination of Martin Luther King. The counterculture had shifted from the flower power of '67 into the politics of struggle of '68. And yet, Lennon's stance here was to reject the wholesale ditching of love and peace in favor of something more "destructive."

When a version of the song was released as the B-side of "Hey Jude" in late August 1968, it wasn't well received by a section of the global left. Lennon had brooded over his political position during the writing and recording of the song, changing his mind about whether he was "in" or "out" of the struggle and, in the end, opting for both, which can be heard on the track, in what Alan Pollack describes as the "typically Lennon-esque ambiguity between tender encouragement and nasty ridicule."

"Honey Pie"	Recorded: October 1, 2, 4, 1968	McCartney *(vocals, piano, lead guitar)*	Lennon *(lead guitar)*
(Lennon-McCartney)	UK Release: November 22, 1968	Harrison *(bass guitar)*	Starr *(drums)*

Additional contributors: Various unnamed musicians

A 1920s-style music hall pastiche, this song showcases McCartney's musical versatility, which no one else in the band could ever hope to match.

"Both John and I had a great love for music hall," McCartney revealed later. "I very much liked that old crooner style, the strange fruity voice that they used, so 'Honey Pie' was me writing one of them to an imaginary woman, across the ocean, on the silver screen, who was called 'Honey Pie.' It's another of my fantasy songs."

With Harrison playing the new six-string Fender bass that had recently been delivered to the band, the song seemingly needed only one take to complete. Lennon added a jazzy guitar solo which Harrison later recalled as a "a brilliant solo—sounded like Django Reinhardt or something. It was one of them where you just close your eyes and happen to hit all the right notes. Sounded like a little jazz solo."

"Savoy Truffle"	Recorded: October 3, 5, 11, 14, 1968	McCartney *(bass guitar)*	
(Harrison)	UK Release: November 22, 1968	Harrison *(double-tracked vocals, lead guitar)*	Starr *(drums, tambourine)*

Additional contributors: Various unnamed musicians

Written by Harrison as a humorous tribute to Eric Clapton's sweet tooth and fondness for chocolate, the lyrics of "Savoy Truffle" more or less lists all the varieties one would find in a box of British Confectionary Mackintosh's Good News chocolates, including the song's title.

In his autobiography, Harrison explains he got a little help for the bridge lyrics: "I got stuck with the two bridges for a while and Derek Taylor wrote some of the words in the middle—'You know that what

The Beatles (White Album) (1968)

you eat you are,'" derived from the name of the 1968 film *You Are What You Eat*.

Recording took place without Lennon at Trident Studios, with Harrison contributing a stinging guitar solo, which sounds like the shrill of a dentist's drill. Harrison wanted a brass section on the song, and George Martin handed the task of arranging and scoring this to his assistant, Chris Thomas, who found it a "real chore." As engineer Brian Gibson explained to Mark Lewisohn in *The Complete Beatles Recording Sessions*, "The session men were playing really well – there's nothing like a good brass section letting rip—and it sounded fantastic. But having got this really nice sound, George turned to Ken Scott and said, 'Right, I want to distort it.' The musicians came up to the control room to listen to a playback and George said to them, 'Before you listen I've got to apologize for what I've done to your beautiful sound. Please forgive me—but it's the way I want it!'"

"Savoy Truffle" is one of two *White Album* songs to refer to other Beatles tracks—the other being "Glass Onion."

Cry Baby Cry	Recorded: July 15, 16, 18, 1968	McCartney *(bass guitar)*	Lennon *(vocals, acoustic guitar, piano, organ)*
(Lennon-McCartney)	UK Release: 22 November 1968	Harrison *(lead guitar)*	Starr *(drums, tambourine)*

Additional contributors: George Martin (harmonium)

Compared to the "hush little baby" and "once upon a time" world of the nursery rhyme, Lennon's lullaby "Cry Baby Cry" is dark indeed. He was drawing on a strong tradition in British culture. It tacitly references everything from "Ring a Ring o' Roses," the cryptic rhyme about the deadly plague that killed millions of people in medieval Europe ("A-tishoo! A-tishoo! We all fall down"), through "Goosey Goosey Gander," a rhyme about witch-hunting Catholics at a time of religious persecution ("There I met an old man, who wouldn't say his prayers, so I took him by his left leg and threw him down the stairs"), to "Mary Mary Quite Contrary," a rhyme about Bloody Mary, Henry VIII's daughter, that references graveyards which were filling with martyred Protestants under her reign, with the "silver bells" in the rhyme representing thumbscrews and "cockleshells," which were also medieval instruments of torture that were attached to male genitals. All very dark indeed. Lennon's lullaby centers around a descending sequence in one chord (E minor) and is an evocative piece with lyrics like "she's old enough to know better," a "seance in the dark," and its "voices out of nowhere/Put on specially by the children for a lark."

The song is supposedly based on a television advert, with the lyric "Cry baby cry, make your mother buy" later developing into the darker nursery rhyme. Structurally, "Cry Baby Cry" oozes the typically creepy and surreal atmosphere of a standard old nursery rhyme, both by the verbal ambiguity of the lyrics, with their alternating King and Queen, Duchess and Duke, and also by an ominously recurring blues feel to the backing track.

Of the four Beatles, Lennon was often the most in touch with his lost childhood (the outro to "Cry Baby Cry" repeats the refrain, "Can you take me back where I came from? Can you take me back?") and this song, with its chimerical now sunlit, now darkened, corridors, is one of his most haunting compositions.

"Revolution 9"	Recorded: June 6/10/11/20/21, 1968	Collage of sound	
(Lennon-McCartney)	UK Release: November 22, 1968		

In the early days of "Love Me Do" and "Please Please Me," track length was driven to brevity by a desire to get the best audio quality possible. George Martin and The Beatles understood this science of audio perfectly well. On their first seven albums, the three-minute song barrier was breached (albeit marginally) only twice; "You Really Got A Hold On Me" came in at 3:01, and "Ticket To Ride" at 3:09. Now consider "Revolution 9."

The Beatles (White Album) (1968)

Coming in at 8:22, it's not only the band's longest "song," it also stretched the limits of what a song could be. An overwhelming assault of sound, the track borrows the experimental style first toyed with on "Tomorrow Never Knows" and takes it to the extreme.

This sonic collage is comprised of old-school instruments, such as oompah brass and percussion, and mashes them up with modern techniques such as looping tape samples and mellotron, as well as sound and studio effects. Layered upon all this is the human voice in the form of laughter and speeches, mumbling and screaming, and various other vocal fragments from Lennon, Harrison, Yoko, George Martin, and Alistair Taylor, assistant of Brian Epstein.

"Revolution 9" is the world's most widely distributed avant-garde artifact. Approximately one million households owned a copy of the work within mere days of its release and, sixty years later, its listeners number in the many hundreds of millions. It remains, no matter how uncomfortable for the listener, an attempt by Lennon to present on record the unconscious elements of his creativity. "Revolution 9" is, in the random juxtaposition of a sound collage, a representation of the semi-asleep, media-hopping state of mind Lennon liked to relax in.

"Good Night"	Recorded: June 28, July 2/22, 1968		George Martin *(celesta)*
(Lennon-McCartney)	UK Release: November 22, 1968		Starr *(vocals)*

Additional contributors: George Martin (celesta) and various unnamed musicians

A tender lullaby written for his son Julian, the song was so out of character for Lennon that many believed it was written by McCartney. Lennon gave the song for Ringo to sing, most likely because it better suited his melancholy delivery, and they originally recorded it first with Lennon on acoustic guitar, and then later on piano.

A few weeks later, all this work was scrapped in favor of an orchestral arrangement with choir, both scored by George Martin, which Lennon later described as "possibly overlush."

Years later, McCartney remembered this song fondly.

"John rarely showed his tender side, but my key memories of John are when he was tender, that's what has remained with me; those moments where he showed himself to be a very generous, loving person. I always cite that song as an example of the John beneath the surface that we only saw occasionally."

Yellow Submarine (1969)
Full Steam Ahead

"John Lassester, the director of Toy Story *and former chief creative officer of Pixar and Walt Disney Animation, has called* Yellow Submarine *a 'revolutionary work' that helped 'pave the way for the fantastically diverse world of animation that we all enjoy today.' Josh Weinstein, writer for* The Simpsons, *has claimed the film 'gave birth to modern animation,' and that without it—and its subversive humor—we'd have no* South Park, Toy Story, *or* Shrek.*"*
—Holly Williams, *BBC Culture* (2018)

Yellow Submarine	Released: January 13, 1969	Recorded: May 26, 1966– February 11, 1968 (The Beatles) October 22-23, 1968 (George Martin)	Duration: 39:16
Producer: George Martin	Studio: EMI and De Lane Lea, London	Label: Apple	Tracks: 13

Track Listing

Side One

No.	Title	Lead Vocals	Length
1	"Yellow Submarine"	Starr	2:39
2	"Only A Northern Song"	Harrison	3:24
3	"All Together Now"	McCartney	2:11
4	"Hey Bulldog"	Lennon	3:12
5	"It's All Too Much"	Harrison	6:26
6	"All You Need Is Love"	Lennon	3:47

Side Two

No.	Title	Lead Vocals	Length
7	"Pepperland"	Instrumental	2:18
8	"Sea Of Time"	Instrumental	3:00

(Continued)

Yellow Submarine (1969)

9	"Sea Of Holes"	Instrumental	2:16
10	"Sea Of Monsters"	Instrumental	3:35
11	"March Of The Meanies"	Instrumental	2:16
12	"Pepperland Laid Waste"	Instrumental	2:09
13	"Yellow Submarine In Pepperland"	Instrumental	2:10

All songs written by Lennon-McCartney, except tracks 2 and 5, written by Harrison, and tracks 8-12, written by George Martin

Track by Track: *Yellow Submarine*

"Yellow Submarine"	Recorded: May 25; June 1, 1966	McCartney *(backing vocals, bass guitar, shouting)*	Lennon *(backing vocals, acoustic guitar, shouting)*
(Lennon-McCartney)	UK Release: August 5, 1966	Harrison *(backing vocals, tambourine)*	Starr *(vocals, drums)*

See under previous entry for *Revolver* on page 118.

"Only A Northern Song"	Recorded: February 13-14; April 20, 1967	McCartney *(bass guitar, trumpet, tape-effects, noises)*	Lennon *(piano, glockenspiel, tape-effects, noises)*
(Harrison)	UK Release: January 17, 1969	Harrison *(vocals, organ, tape-effects, noises)*	Starr *(drums)*

By 1966, George Harrison had become a little jaded and lost interest in music. His time spent in India later that year had awakened the spiritual side of him that he found hard to reconcile with his life as a rock star.

"Only A Northern Song" was recorded during the *Sgt. Pepper* sessions but, understandably, not deemed good enough to make it onto the album. His contribution to the album was sparse to say the least. The target of the song itself was the publishing company Northern Songs Ltd, set up by Dick James to publish the Lennon-McCartney songs. Harrison and Starr were effectively contracted as songwriters, which irked Harrison. As Harrison himself commented, "The song was copyrighted by Northern Songs Ltd, which I don't own, so: 'It doesn't really matter what chords I play . . . as it's only a Northern Song.'"

The track itself is quite psychedelic, containing a variety of instruments such as a glockenspiel played by Lennon while McCartney adds a trumpet part. Like his song "Blue Jay Way," this song was composed on a Hammond B3 organ, which was quite common for Harrison during this mid-Beatles period. Lyrical references such as "chords are wrong" or "out of key" display its writer's sarcasm and enhances the sense of dissonance in the track. Even though the song was rejected for *Sgt. Pepper*, it was the first song to be offered by the band for the soundtrack album of their animated film *Yellow Submarine*.

"All Together Now"	Recorded: May 12, 1967	McCartney *(vocals, bass guitar, acoustic guitar, claps)*	Lennon *(backing vocals, acoustic guitar, ukelele, harmonica, claps)*
(Lennon-McCartney)	UK Release: January 17, 1969	Harrison *(backing vocals, claps)*	Starr *(drums, finger cymbals, claps)*

Written in the studio for the *Yellow Submarine* movie, "All Together Now" is a childlike singalong written in the music hall tradition. McCartney conjured up the main sections of the song, while Lennon contributed the "sail the ship, chop the tree" middle section. As McCartney explains in *Many Years From Now*, "It's really a children's song. I had a few young relatives and I would sing songs for them. I used to do a song for kids called 'Jumping Round The Room,' very similar to 'All Together Now,' and then it would be 'lying on your backs,' all the kids would have to lie down, then it would be 'skipping round the room,' 'jumping in the air.' It's a play away command song for children." Nonetheless, McCartney was delighted when the song became a popular terrace chant at soccer matches shortly after its release in early 1969.

Yellow Submarine (1969)

"Hey Bulldog"	Recorded: February 11, 1968	McCartney *(harmony vocals, bass guitar)*	Lennon *(double-tracked vocals, piano, lead guitar)*
(Lennon-McCartney)	UK Release: January 17, 1969	Harrison *(guitar)*	Starr *(drums, tambourine)*

Having met to film a promotional video for "Lady Madonna," the band found themselves in a particularly productive mood. Easily the best song on this album which most neglect to include in their collections, "Hey Bulldog" has since enjoyed a degree of cult status, perhaps helped by the song's inadvertent relative anonymity. Lennon had the vague idea of a song which he'd called "Hey Bullfrog" and, on showing it to his songwriting partner, the pair polished it off in a day by recording it for *Yellow Submarine*.

A rawly exciting track, now kick-ass, now edgy, "Hey Bulldog," based on a bluesy variation of Berry Gordy's "Money," musically chimes with its uncertain and ambiguous lyric (as Alan Pollack wittily points out, given Lennon's scathing delivery in the vein of "I Am The Walrus": "Do you really believe the protagonist is interested in talking to you if you're lonely?").

Like other sterling compositions by the band, "Hey Bulldog" grips the listener until the very final moments, with an outro that becomes a focal point of the track by redefining the earlier parts of the song through the use of sound effects and typical Beatles goofing about. The song was later said by engineer Geoff Emerick to be one of their final true band efforts, with equal contributions from all members.

"It's All Too Much"	Recorded: May 25-26; June 2, 1967	McCartney *(harmony vocals, bass guitar)*	Lennon *(harmony vocals, lead guitar)*
(Harrison)	UK Release: January 17, 1969	Harrison *(double-tracked vocals, lead guitar, organ)*	Starr *(drums, tambourine)*

Additional contributors: Various unnamed musicians

Another one of Harrison's songs composed on a Hammond B3 organ, "It's All Too Much" was written following some of his experiences with LSD.

"I just wanted to write a rock 'n' roll song about the whole psychedelic thing of the time," George wrote in his autobiography, *I Me Mine*. "Because you'd trip out, you see, on all this stuff, and then whoops! You'd just be back having your evening cup of tea!"

Harrison also revealed that he drew parallels between the "realizations" brought about by his LSD experiences and his experiences with Transcendental Meditation. Recorded a month after completing the *Sgt. Pepper* album, the track is a rare example of a mainly self-produced track by the band. A psychedelic song with a heavy Indian influence, the lyrics don't merely reflect the optimism of the Summer of Love period.

Several Beatles biographers understandably dismiss the track as aimless.

"I have no doubt of George's mystical sincerity," Alan Pollack stated. "But I cannot escape the feeling in this song that he is self-effacingly winking at us with the desideratum, 'Show me that I'm everywhere and get me home in time for tea.' It's an obliquely phrased mixed message reflecting of an inner conflict between spiritual striving and the backsliding lust for bourgeois comfort and respectability."

"All You Need Is Love"	Recorded: June 14, 19, 23-25, 1967	McCartney *(harmony vocals, string bass, bass guitar)*	Lennon *(vocals, harpsichord, banjo)*
(Lennon-McCartney)	UK Release: July 7, 1967	Harrison *(harmony vocals, violin, guitar)*	Starr *(drums)*

Additional contributors: Various unnamed musicians

See under previous entry for "Magical Mystery Tour" on page 175.

ABBEY ROAD (1969)
And in the End

"After Let It Be *I really thought we were finished. So I was quite surprised when Paul rang up and said, 'Look, you know, what happened to* Let It Be *is silly. Let's try to make a record like we used to. Would you come and produce it like you used to?' I said, 'Well, I'll produce it like I used to if you'll let me.' So Paul rounded up John, George and Ringo and we started work on* Abbey Road. *It really was very happy, very pleasant, and it went frightfully well."*
—George Martin, *All You Need Is Ears* (1994)

Abbey Road	Released: September 26, 1969	Recorded: February 22–August 20, 1969	Duration: 47:03
Producer: George Martin	Studio: EMI, Olympic and Trident, London	Label: Apple	Tracks: 17

Track Listing

Side One

No.	Title	Lead Vocals	Length
1	"Come Together"	Lennon	4:19
2	"Something"	Harrison	3:02
3	"Maxwell's Silver Hammer"	McCartney	3:27
4	"Oh! Darling"	McCartney	3:27
5	"Octopus's Garden"	Starr	2:51
6	I Want You (She's So Heavy)	Lennon	7:47

Side Two

No.	Title	Lead Vocals	Length
7	"Here Comes The Sun"	Harrison	3:05
8	"Because"	Lennon, McCartney, Harrison	2:45
9	"You Never Give Me Your Money"	McCartney	4:03
10	"Sun King"	Lennon	2:26

(Continued)

Abbey Road (1969)

11	"Mean Mr. Mustard"	Lennon	1:06
12	"Polythene Pam"	Lennon	1:13
13	"She Came In Through The Bathroom Window"	McCartney	1:58
14	"Golden Slumbers"	McCartney	1:31
15	"Carry That Weight"	McCartney	1:36
16	"The End"	McCartney	2:05
17	"Her Majesty"	McCartney	0:23

All songs written by Lennon-McCartney, except tracks 2 and 7, written by Harrison, and track 5, written by Richard Starkey

And in the End

A matter of days after the famous rooftop concert, the band reconvened at Trident on February 22 to begin work on *Abbey Road*. Regularly cited as one of the greatest, if not *the* greatest rock album of all time (in recent years eclipsing *Sgt. Pepper* in this regard), *Abbey Road* is The Beatles' most technically accomplished album.

After the *Get Back* sessions, McCartney had suggested to George Martin that the band make an album the way "we used to do it." Martin agreed on the strict condition that all the band, especially Lennon, allow him to produce the record in the same manner as earlier albums. And so, with a disciplined George Martin back at the desk, the songs feature eight-track recordings replete with smooth, pellucid sounds: the pseudo-velvety textures of the Moog synthesizer, a richly deep bass end, Harrison's ubiquitous Leslie-toned guitar, and with Ringo's always invaluable contribution being captured more roaringly than ever. Once more, the band tries on different musical genres.

The album incorporates styles such as pop, rock, and blues, as well as progressive rock, and is notable for having a "Long Medley" of songs on the second side which has been covered as one suite by other prominent artists. *Abbey Road* is the last album the band *recorded*, although "Let It Be" was the last album *released* before the band's break-up in April of 1970.

Track by Track: *Abbey Road*

"Come Together"	Recorded: July 21, 22, 23, 25, 29, 30, 1969	McCartney *(harmony vocals, bass guitar, electric piano)*	Lennon *(vocals, rhythm guitar, lead guitar, claps)*
(Lennon-McCartney)	UK Release: September 26, 1969	Harrison *(guitar)*	Starr *(drums, maracas)*

Early in his post-Beatles solo career, Lennon championed causes such as "Give Peace A Chance," "Power To The People," and "Imagine." "Come Together" is the last of Lennon's counterculture declarations while still in the band. The track was conceived by Lennon as a political rallying cry for the writer and pro-drugs activist, Timothy Leary, and his campaign to stand against Ronald Reagan as governor of California.

With its deliberately sexual title, the song is an anti-elitist call to free the imagination from all fetters, political and emotional. After the first two verses, with their stream of self-confessed and menacing "mumbo-jumbo" lampooning the hippie figureheads who seek followers among the dropouts of society, the song resolves into the impactful clarity of "One thing I can tell you is you got to be free."

Ian MacDonald, in his book *Revolution in the Head* suggests, "Nothing else on *Abbey Road* matches the Zeitgeist-catching impact" of "Come Together," the track being "the key song of the turn of the decade," adopting the "then-new American 'laid-back' or 'spaced-out' style, in which a stoned laziness of beat and a generally low-profile approach offered a cool proletarian alternative to middle-class psychedelic artifice."

Around Lennon's galvanizing vocal, the other Beatles prowl in the most organically stranger-than-life and weird way (in the words of McCartney, it was a "swampy" groove); the drums are smoother than ever, the brilliantly idiomatic bass is vibrantly resonant, and the double-tracked guitar solo simply smokes. As Lennon confessed in *Anthology*, the band sensed what the track needed musically "because we've played together a long time."

Abbey Road (1969)

"Something"	Recorded: April 16; May 2, 5; July 11/16; August 15, 1969	McCartney *(backing vocals, bass guitar, claps)*	Lennon *(guitar)*
(Harrison)	UK Release: September 26, 1969	Harrison *(double-tracked vocals, lead guitar, claps)*	Starr *(drums, claps)*

Additional contributors: Billy Preston (organ)

"Something," George Harrison's Magnum Opus, was once described by Frank Sinatra as the "the greatest love song of the last fifty years." The fact that he also described it as the best Lennon-McCartney love song was the highest praise indeed. As mentioned in previous sections, Harrison had always suffered in the shadow of his two illustrious bandmates and had struggled to get his songs onto their albums.

Assumed to be written for his wife, Pattie Boyd, Harrison later elaborated in a *Rolling Stone* interview in 1976 that it was written about his connection to the Hindu deity, Krishna, "All love is part of a universal love," Harrison explained. "When you love a woman, it's the God in her that you see."

This stunningly beautiful ballad started life on a piano in an empty Abbey Road studio during the sessions for the *White Album*. Unsure whether his melody had been "borrowed" subconsciously from another song, he set it aside. Harrison came back to it during the *Get Back* sessions in January 1969, though he was still uncertain about his lyrics and asked Lennon for help. Over the space of several sessions, the band laid down the main track with the addition of organ from Billy Preston. The two instrumental highlights on the track are McCartney's wonderfully fluid bassline and Harrison's tasteful guitar solo.

According to Geoff Emerick, Harrison was not happy with McCartney's original bass part.

"Paul started playing a bassline that was a little elaborate, and George told him, 'No, I want it simple.' Paul complied. There wasn't any disagreement about it, but I did think that such a thing would never have happened in years past. George telling Paul how to play the bass? Unthinkable! But this was George's baby, and everybody knew it was an instant classic."

Lennon described it as "about the best track on the album, actually," while McCartney added, "For me, I think it's the best he's written," while George Martin confessed, "I was surprised he had it in him!" "Something" marked the first time a George Harrison song was the A-side of a Beatles single when it was released in October 1969, and its popularity has endured over the years and remains the second most-covered Beatles song after "Yesterday." Harrison had indeed finally come out of the shadow to compose a song equal in stature to his two illustrious band mates.

"Maxwell's Silver Hammer"	Recorded: July 9-11; August 6, 1969	McCartney *(vocals, backing vocals, guitars, piano, Moog synthesizer)*	
(Lennon-McCartney)	UK Release: September 26, 1969	Harrison *(backing vocals, lead guitar)*	Starr *(drums, backing vocals, anvil)*

Additional contributors: George Martin (organ)

In January of 1966, while on his way home to Liverpool in his Aston Martin, McCartney was listening to BBC Radio 3 and came across a version of the play *Ubu Cocu* (Ubu Cuckolded) by French symbolist writer Alfred Jarry. Described in the *Radio Times* as a "pata-physical extravaganza," the play made a deep impression on McCartney. "Pataphysics" was an invention of Jarry's, intended as a parody of science ("to poke fun at toffee-nosed academics," as McCartney says in *The Lyrics*), and was a sardonic "science of imaginary solutions."

"Maxwell's Silver Hammer," as McCartney points out in *The Lyrics*, also features two prominent scientists in the name of the song's protagonist, Maxwell Edison ("Maxwell" a descendant of James Clerk Maxwell, a pioneer of electromagnetism, and "Edison" after Thomas Edison, an inventor connected to the lightbulb and the phonograph, which is very apt considering the protagonist is making an appearance on a gramophone record.)

Abbey Road (1969)

Considered by its composer to be a potential Beatles single, and invoking the world of the children's nursery rhyme, "Maxwell's Silver Hammer" is a song about a serial killer who uses a silver hammer to clobber his victims to death. The recording of the track was also not without its victims, as McCartney confesses: "I was very keen on this song, but it took a bit long to record, and the rest of the guys were getting pissed with me."

Although McCartney also makes the interesting general point about recording at that time, "Recording sessions were always good because no matter what our personal troubles were, no matter what was happening on the business front, the minute we sat down to make a song, we were in good shape. Right until the end, there was always a great joy in working together in the studio."

"Oh! Darling"	Recorded: April 20, 26; July 17, 18, 22; August 11, 1969	McCartney *(vocals, backing vocals, bass guitar, guitar)*	Lennon *(backing vocals, piano)*
(Lennon-McCartney)	UK Release: September 26, 1969	Harrison *(backing vocals, guitar, synthesizer)*	Starr *(drums)*

During the early days in Hamburg, McCartney learnt his screaming voice from Little Richard. Early Beatles covers, like "Long Tall Sally" and "Kansas City," demonstrated how well he had mastered it. But when it came to reviving the technique five years later for the doo-wop style rocker "Oh! Darling," he struggled a little.

According to engineer Alan Parsons, "Paul came in several days running to do the lead vocal on 'Oh! Darling.' He'd come in, sing it and say, 'No, that's not it, I'll try it again tomorrow.' He only tried it once per day, I suppose he wanted to capture a certain rawness which could only be done once before the voice changed. I remember him saying, 'Five years ago I could have done this in a flash.'"

McCartney wanted it to sound as though he'd been performing it on stage all week. Rehearsed first during the *Get Back* sessions in

January 1969, the song was recorded at various sessions up until August, when the exquisite three-part doo-wop vocal harmonies were added. The rawness of the song appealed to Lennon who, typically and immodestly, commented later.

"That was a great one of Paul's that he didn't sing too well. I always thought that I could've done it better—it was more my style than his."

"Octopus's Garden"	Recorded: April 26, 29; July 17, 18, 1969	McCartney *(backing vocals, bass guitar, piano)*	Lennon *(guitar)*
(Starkey)	UK Release: September 26, 1969	Harrison *(backing vocals, lead guitar, synthesizer)*	Starr *(vocals, drums, percussion, effects)*

During the recording of "Back In The U.S.S.R." in late August of 1968, Ringo had temporarily left the band after a row with McCartney over the drum part. As a remedy to the stress, Ringo took his family to the Mediterranean island of Sardinia. While on vacation, Ringo became fascinated with the octopus, that soft-bodied, eight-limbed mollusk, which boasts over three hundred species and can radically alter its shape, with the creature able to squeeze through very small gaps.

The behavioral trait of the octopus which most delighted Ringo was its gardening. A local fisherman informed the famous drummer that the creature liked to roam the seabed, collecting stones and shiny objects, and use them to build gardens. The function of the garden building is to create nesting grounds for their eggs, as a safe space to chill, and as a way to decorate their homes to give it a personal touch. Octopuses have even been known to use decorations outside their hidden gardens which may act as an underwater burglar alarm.

Given the ingenious nature of these creatures, and the undoubted attraction of the tranquility of the gardens versus the fractious air of Abbey Road, Ringo set about celebrating the octopus in song, his second for the band. A commonplace tune, in country-and-western style, it's an account of a Peter Pan-like escape from the harsh reality of the storm, and being told what to do, in a garden that is a place of safety,

Abbey Road (1969)

happiness, and joy. The song was affectionately welcomed back to Abbey Road, along with its composer on his return. We see Ringo working on the song with Harrison in Peter Jackson's documentary *Get Back*.

"I Want You (She's So Heavy)"	Recorded: February 22, April 18, 20, August 8/11, 1969	McCartney *(harmony vocals, bass guitar)*	Lennon *(vocals, harmony vocals, lead guitars, organ, Moog synthesizer)*
(Lennon-McCartney)	UK Release: September 26, 1969	Harrison *(harmony vocals, lead guitars)*	Starr *(drums, congas)*

Additional contributors: Billy Preston (organ)

One of the most complex recordings the band ever committed to tape, "I Want You" is Lennon's obsessive paean to Yoko Ono. As uncompromising as its creator, the song contains just fourteen words, repeated over and over. The obsessiveness of the lyrics is mirrored by the repetitiveness of the music, which is essentially a basic blues riff.

Lennon told *Rolling Stone* magazine in 1970, "When you're drowning, you don't say 'I would be incredibly pleased if someone would have the foresight to notice me drowning and come and help me,' you just scream."

The song was the first to be recorded for *Abbey Road* and one of the last tracks to be completed. The complexity of the recording came from the various takes that were spliced together to form mixes of the song over six months, not to mention the various overdubs.

The ending of the track is truly momentous. The final three minutes of the almost eight-minute song build into a mass of overdubbed guitars, with a slow-building wall of white noise generated by Lennon on the Moog synthesizer that sounds like a howling wind.

"John and George went into the far lefthand corner of studio 2 to overdub those guitars," according to sound engineer Jeff Jarrat. "They wanted a massive sound so they kept tracking and tracking, over and over."

Finally, the track cuts off abruptly.

"When they recorded the backing track," engineer Geoff Emerick explains, "The Beatles had just played on and on, with no definitive conclusion, so I assumed I would be doing a fade-out. John had other ideas, though. He let the tape play until just twenty seconds or so before the take broke down, and then all of a sudden he barked out an order: 'Cut the tape here.' 'Cut the tape?' I asked, astonished. We had never ended a song that way. 'You heard what I said, Geoff; cut the tape.' I glanced over at George Martin, who simply shrugged his shoulders, so I got out the scissors and sliced the tape at precisely the point John indicated. At the time, I thought he was out of his mind, but due to the shock factor, it ended up being incredibly effective, a Lennon concept that really worked."

"Here Comes The Sun"	Recorded: July 7, 8, 16; August 6, 15, 19, 1969	McCartney *(backing vocals, bass guitar, claps)*	
(Harrison)	UK Release: September 26, 1969	Harrison *(vocals, backing vocals, acoustic guitar, harmonium, Moog synthesizer, claps)*	Starr *(drums, claps)*

Additional contributors: Various unnamed musicians

George Harrison did not join The Beatles with the intention of becoming a businessman. But, following the creation of the band's Apple Corps Limited in 1968, he became just that. Harrison recalls, "'Here Comes The Sun' was written at the time when Apple was getting like school, where we had to go and be businessmen; 'sign this' and 'sign that.' So, one day, I decided I was going to sag off Apple and I went over to Eric Clapton's house. The relief of not having to go and see all those dopey accountants was wonderful, and I walked around the garden with one of Eric's acoustic guitars and wrote 'Here Comes The Sun.'" The sight of Harrison in full flow stayed with Clapton, who remembers, that "he was just a magical guy and he would show up, get out of the car with his guitar and come in and start playing . . . I just watched this thing come to life. I felt very proud that it was my garden that was inspiring it."

The song had a significant impact on Harrison's spiritual journey, and marked a turning point in his life.

"I was getting disillusioned with the whole Beatles thing... 'Here Comes The Sun' was a song that came out of that period, and it was a very uplifting and positive song."

The lyrics find Harrison lamenting how long the English winter lasts, greeting the spring sunshine with an open and optimistic heart. Harrison's acoustic guitar part, with its stunning arpeggios, has become a favorite of budding guitarists everywhere, and reflects the joyous liberation of the coming of spring. In keeping with the band's endless search for sonic experimentation, the *Abbey Road* album contains one of the first uses in the UK of the Moog synthesizer, a recent invention in 1969. Harrison had one made especially for him, and uses it here to create subtle, atmospheric effects.

"Here Comes The Sun" arguably stands as Harrison's finest and, in May 2023, it became the first Beatles song to surpass one billion plays on Spotify, the first song from the 1960s to achieve the milestone. Enough said.

"Because"	Recorded: August 1, 4, 5, 1969	McCartney *(vocals, bass guitar)*	Lennon *(vocals, lead guitar)*
(Lennon-McCartney)	UK Release: September 26, 1969	Harrison *(vocals, Moog synthesizer)*	

Additional contributors: George Martin (harpsichord)

The final song recorded for *Abbey Road*, Lennon's "Because," is a three-part harmony inspired by hearing Yoko Ono play Beethoven's "Moonlight Sonata." As George Martin has overdubbed the track twice, the song boasts nine voices in all, the most complex harmonies on any of the band's records.

Understandably, even under the tutelage and expert ear of a producer the caliber of Martin, the band took some time before "Because" hit perfection. The vocals create a sumptuous mood of rapt contemplation which, rather than engaging the heart, seems to float along on the hot air of its high-flown wordplay. As a result, "Because" has a somewhat classic and transcendent detachment, helped along by the chordal melody of the backing from harpsichord, guitar, and synthesizer.

"You Never Give Me Your Money"	Recorded: May 6; July 1, 15, 30, 31; August 5, 1969	McCartney *(multi-tracked vocals, backing vocals, bass guitar, guitars, pianos, wind chimes, tape-loops)*	Lennon *(backing vocals, guitars)*
(Lennon-McCartney)	UK Release: September 26, 1969	Harrison *(backing vocals, guitars)*	Starr *(drums, tambourine)*

McCartney's poignant lament about the band's business battles of early 1969, "You Never Give Me Your Money" shaped the Long Medley that reigns over the second half of the album. In McCartney's 1969 notebook, three separate songs are listed, "You Never Give Me Your Money," "Out Of College," and "One Sweet Dream."

In the words of Ian MacDonald, "The Beatles' future may be gone, but McCartney is determined to salvage their spirit, and that of the Sixties, for *his* future. 'You Never Give Me Your Money' marks the psychological opening of his solo career."

McCartney explained in *The Lyrics*, "The Beatles stuff all got too heavy. . . . It meant having to go into meetings and sit in the boardroom with all the other Beatles and with the accountants and with this guy Allen Klein . . . I smelled a rat but the other chaps didn't, so we had a fight over it and I got voted down . . . The Beatles were already beginning to break up. John had said he was leaving, and Allen Klein told us not to tell anyone. . . . We were living a lie. . . . And we all knew that phase of our lives, of being The Beatles, was coming to an end."

Abbey Road (1969)

"Sun King/Mean Mr. Mustard"	Recorded: July 24, 25, 29, 1969	McCartney *(harmony vocals, bass guitar, harmonium, piano, tape-loops)*	Lennon *(multi-tracked vocals, lead guitar, maracas)*
(Lennon-McCartney)	UK Release: September 26, 1969	Harrison *(lead guitar)*	Starr *(drums, tambourine, bongos)*

Additional contributors: George Martin (organ)

Lennon's abiding love of wordplay had not only surfaced many times in the band's recordings, but also in his modest literary career with the publication of his award-winning books, *In His Own Write* and *A Spaniard in the Works*. "Sun King" boasts a mock Spanish/Italian/Portuguese wordplay and is quite unique, even for The Beatles. The reader might be aware of Lennon's pigeon German in the bathtub scene of the *A Hard Days Night* movie, and his mumbling nonsense at the end of "I'm So Tired."

Here, against a backdrop of sumptuous mock Mediterranean guitar stolen from Fleetwood Mac's "Albatross," the phrases ("Quando para mucho mi amore de felice corazon/Mundo paparazzi mi amore chica ferdi parasol") work well against the music, suggesting Lennon had an excellent ear for the lilt and rhythm of the languages. The song segues from this stress-busting siesta of "Sun King" into the mischievous "Mean Mr. Mustard," while keeping the same tempo.

The song was based on the miserly sixty-five-year-old Scotsman John Alexander Mustard, whom Lennon had read about in the *Daily Mirror* back in June of 1967. Poor Mustard, who not only shaved and went to bed in the dark but also hid five-pound notes, had been taken to a divorce court by his wife due to his meanness.

"Polythene Pam/She Came In Through The Bathroom Window"	Recorded: July 25, 28, 30, 1969	McCartney *(double-tracked vocals, backing vocals, bass guitar, lead guitar, piano, electric piano)*	Lennon *(double-tracked vocals, backing vocals, twelve-string acoustic guitar)*
(Lennon-McCartney)	UK Release: September 26, 1969	Harrison *(backing vocals, lead guitar)*	Starr *(drums, tambourine, maracas, cowbell)*

Female fans are the theme of this compositional couplet. Lennon's "Polythene Pam" is rousingly sung in an outrageous Scouse accent for earthy authenticity, a clue that this song harks back to the days of The Cavern and one "Polythene Pat," an early fan, and real-life Liverpool "Judy," who was well known for her plastic fetish.

Powered throughout by a delicious pseudo-samba drive, the song starts with meaty-sounding power chords on a twelve-string acoustic and even resurrects the band's "yeah, yeah, yeah" refrain from the days of The Cavern and "She Loves You," before segueing into McCartney's "She Came In Through The Bathroom Window."

This second semi-track is a song about the so-called "Apple scruffs," the female fans who used to hang around Apple, Abbey Road, and the band's private houses. As legend has it, one such scruff managed to climb a ladder left in McCartney's garden, stole into his home through the bathroom window (naturally), and purloined a precious photo of his father. McCartney then had to negotiate with the scruffs for the safe return of the photograph. (Lennon, meanwhile, claims the song is about Linda McCartney but also confesses, "I don't know; *somebody* came in the window!")

As with all the songs in *Abbey Road*'s Long Medley, "She Came In Through The Bathroom Window" is highly desirable rock. It boasts a seductively creative thread of phrases on guitar, probably by McCartney, coupled with a walking bass and a powerhouse percussion which drives the rhythmic feel of the track.

Abbey Road (1969)

"Golden Slumbers/ Carry That Weight"	Recorded: July 2, 3, 4, 30, 31; August 15, 1969	McCartney *(double-tracked vocals, chorus vocals, rhythm guitar, piano)*	Lennon *(chorus vocals)*
(Lennon-McCartney)	UK Release: September 26, 1969	Harrison *(chorus vocals, bass guitar, lead guitar)*	Starr *(drums, chorus vocals)*

Additional contributors: Various unnamed musicians

"Golden slumbers kiss your eyes / Smiles awake you when you rise / Sleep, pretty wantons, do not cry / And I will sing a lullaby." So begins the original verse of "Golden Slumbers" by Elizabethan poet and dramatist Thomas Dekker (1570–1632).

The track marks the start of the closing sequence of the Long Medley. As McCartney recalled in *Anthology*, "I was playing the piano in Liverpool in my dad's house . . . and I came to 'Golden Slumbers' . . . I couldn't remember the old tune, so I just started playing my own tune to it. I liked the words so I kept them, and it fitted with another bit of song I had," which suggests he had already written "Carry That Weight" and composed the tune for "Golden Slumbers" to echo it.

The recording for this couplet began while Lennon and Yoko were in the hospital recovering from a car accident in Scotland. The lyrics to "Carry That Weight" are truly prescient and poignant. McCartney appears to be saying to the other members of the band, no matter what they do as individuals after The Beatles, it'll never get anywhere near what they did together. He was not wrong. The watching world has always looked back nostalgically to the band's glory days as The Fab Four, and each band member has carried the weight of that achievement for the rest of their lives.

"The End"	Recorded: July 23; August 5, 7, 8, 15, 18, 1969	McCartney *(vocals, backing vocals, bass guitar, piano, lead guitar)*	Lennon *(backing vocals, lead guitar)*
(Lennon-McCartney)	UK Release: September 26, 1969	Harrison *(backing vocals, rhythm guitar, lead guitar)*	Starr *(drums)*

Additional contributors: Various unnamed musicians

The album's Long Medley is brought to a finale with the exhilarating rock 'n' roll climax that is "The End," giving *Abbey Road* arguably the single grandest musical crescendo on any Beatles album, save "A Day In The Life" on *Sgt. Pepper*.

We begin with the band's one and only "drum solo," followed by some power chords and three two-bar guitar solos in sequence from the soaring melodic tones of McCartney, through the Clapton-esque phrasing of Harrison, to the rhythmic grunge of Lennon. The poignancy of "Carry That Weight" is now underlined further by McCartney lyrically returning to the theme of "You Never Give Me Your Money" and the mic-drop line "And in the end / The love you take / Is equal to the love you make."

Musicologist Alan Pollack makes the interesting point that "The End" "is strangely reminiscent of [the work of] composer . . . John Williams, as his signature way of rolling the credits over a musical backing of ultimate, happy ending, philharmonic triumph." In this way, Pollack suggests, *Abbey Road* ends with "the dramatic ethos of . . . the curtain closing number [as] each of the lead players . . . [takes] his curtain call . . . while the audience applauds its head off."

"Her Majesty"	Recorded: July 2, 1969	McCartney *(vocals, acoustic guitar)*	
(Lennon-McCartney)	UK Release: September 26, 1969		

The band had a two-month summer break in the recording of *Abbey Road*, taking off most of May and all of June. Naturally, the workaholic

McCartney was the first back in the saddle after the vacation, laying down this throwaway of a track in the early afternoon before the other Beatles showed up.

An outtake of the song from the January 1969 *Get Back* sessions at Apple records shows "Her Majesty" was already reasonably well worked out months before. McCartney had originally intended to include the song in the album's Long Medley between the two Lennon songs, "Mean Mr. Mustard" and "Polythene Pam."

However, on hearing the playback, McCartney thought the running order didn't pass muster and planned to bin the song completely. Luckily, one of the studio's engineers, John Kurlander, was on strict instructions by the studio to *never* discard anything the band recorded. So, Kurlander edited "Her Majesty" for inclusion at the end of the Medley, detaching it from "The End" with almost half a minute of leader-tape.

By the end of the 1960s, however, Beatles fans were used to such stunts: For example, the outros of both "Hello Goodbye" and "Strawberry Fields Forever" fade out and fade back in. In the case of "Her Majesty," just when the less clued-up fan might think the last recorded Beatles album is finally over, heaving a deep sigh in response to the climax of "The End," they're greeted with the aural shock of a crashing chord and the charming little ditty of "Her Majesty."

Apparently, when McCartney heard Kurlander's edit, he adored the random impact so much, coupled with the fact that the song has already come and gone before you know it, he changed his mind on the track's inclusion. (Incidentally, that crash of a chord at the start of "Her Majesty" is actually the last chord of "Mean Mr. Mustard," a legacy of that track's original link to "Her Majesty" and snipped out of the July 30 mix.)

LET IT BE (1970)
"Get Back" to Basics

"The Beatles are the only group I can think of in rock 'n' roll history that improved to such heights from their early days. It was incredible the way they kept improving; it was like an avalanche. I don't see how anybody could not have been touched by something that they did, even if they were cynical. From all the songs they did, from their beginnings right up through the end, there's no way that you could not have been touched by that."

—Jimmy Page, *Rolling Stone* (2012)

Let It Be	Released: May 8, 1970	Recorded: February 4, 8, 1968; January 24–31, 1969; January 3, 4, 8, 1970; April 1, 1970	Duration: 35:10
Producer: Phil Spector	Studio: Apple, EMI and Olympic Sound, London	Label: Apple	Tracks: 12

Track Listing

Side One

No.	Title	Lead Vocals	Length
1	"Two Of Us"	McCartney and Lennon	3:36
2	"Dig A Pony"	Lennon	3:54
3	"Across The Universe"	Lennon	3:48
4	"I Me Mine"	Harrison	2:26
5	"Dig It"	Lennon	0:50
6	"Let It Be"	McCartney	4:03
7	"Maggie Mae"	Lennon and McCartney	0:40

Side Two

| 8 | "I've Got A Feeling" | McCartney and Lennon | 3:37 |

(Continued)

9	"One After 909"	Lennon with McCartney	2:54
10	"The Long And Winding Road"	McCartney	3:38
11	"For You Blue"	Harrison	2:32
12	"Get Back"	McCartney	3:09

All songs written by Lennon-McCartney, except tracks 4 and 11, written by Harrison, and track 7, traditional

Track by Track: *Let It Be*

| "Two Of Us" | Recorded: January 24, 25, 31, 1969 | McCartney *(vocals, acoustic guitar)* | Lennon *(vocals, acoustic guitar)* |
| (Lennon-McCartney) | UK Release: May 8, 1970 | Harrison *(lead guitar)* | Starr *(drums)* |

Unlike other nostalgic lookbacks at this time, "Two Of Us," working title "On Our Way Home," was a brand-new McCartney composition. Ostensibly about the growing relationship between Paul and Linda, the lyric switches to being about Paul and John, particularly "chasing paper" and "getting nowhere," a reference to the band's contemporary contractual squabbles, which developed into a lawsuit a mere two days after the end of the sessions.

The track's harmonies reminded Lennon and McCartney of their early teenage Everly Brothers infatuation as, during work on the song, they burst into a rendition of "Bye Bye Love."

McCartney explains in *The Lyrics* that he doesn't know where the line "spending someone's hard-earned pay" came from, or what it means. He also adds that he doesn't necessarily *want* meaning, "I don't root for meaning all the time. Sometimes it just feels right. I talked to Allen Ginsberg once about poetry and songs, and Allen told me about a conversation he'd had with Bob Dylan when he was trying to correct Dylan about the grammar of a lyric, and Dylan had said, 'This is a song; it's not a poem.' I know exactly what he means. Sometimes something just sings well. Take 'Spending someone's hard-earned pay.' It can't be 'spending someone's weekly pay packet.' You'd trip yourself up with those words."

"Dig A Pony"	Recorded: January 22, 24, 28, 30; February 5, 1969	McCartney *(harmony vocals, bass guitar)*	Lennon *(vocals, lead guitar*
(Lennon-McCartney)	UK Release: May 8, 1970	Harrison *(lead guitar)*	Starr *(drums)*

In 1980, Lennon dismissed this song as "another piece of garbage." Not one to hold back on his own limitations, he continued, "I was just having fun with words. It was literally a nonsense song. You just take words and you stick them together, and you see if they have any meaning. Some of them do and some of them don't."

With its original title of "All I Want Is You," the song is clearly inspired by his muse, Yoko Ono. Rehearsed throughout January 1969, it was performed as part of their famous Apple rooftop lunchtime concert later that month. A real ensemble performance which they seemed to enjoy, it is this version that was released on the *Let It Be* album.

"Across The Universe"	Recorded: February 4, 8, 1968	McCartney *(backing vocals, piano)*	Lennon *(vocals, backing vocals, acoustic rhythm guitar, lead guitar)*
(Lennon-McCartney)	UK Release: December 12, 1969	Harrison *(backing vocals, sitar, tambura)*	Starr *(maracas)*

Additional contributors: George Martin (Hammond organ) and various other musicians, strings, and choir

Written by Lennon in 1967 during his acid phase, he was unhappy that no one else in the band took the song seriously enough, even going as far as to accuse McCartney of "subconscious sabotage." As he recalled later, "It was a lousy track of a great song and I was so disappointed by it. It never went out as The Beatles. The guitars are out of tune and I'm singing out of tune 'cause I'm psychologically

Let It Be (1970)

destroyed and nobody's supporting me or helping me with it and the song was never done properly."

Lennon thought very highly of the song, telling *Rolling Stone* magazine, "It's one of the best lyrics I've written. In fact, it could be the best. It's good poetry, or whatever you call it." The song is another of Lennon's songs inspired by Maharishi Mahesh Yogi. The phrase *"Jai guru deva, om"* in the chorus is a Sanskrit phrase used by Maharishi that means "Victory to God divine."

"I Me Mine"	Recorded: January 3; April 1, 2, 1970	McCartney *(harmony vocals, bass guitar, organ, electric piano)*	
(Harrison)	UK Release: May 8, 1970	Harrison *vocals, harmony vocals, acoustic guitars, lead guitar*	Starr *(drums)*

Additional contributors: Various unnamed musicians

A "heavy waltz" written by Harrison, "I Me Mine" was the last song to be recorded by the band. The song was written following an experience of selflessness and the ego while on LSD. In his autobiography named after the song, Harrison revealed, "The LSD experience was the biggest experience that I'd had up until that time.... Suddenly I looked around and everything I could see was relative to my ego. I hated everything about my ego. But later, I learned from it, to realize that there is somebody else in here apart from old blabbermouth. The truth within us has to be realized. When you realize that, everything else that you see and do and touch and smell isn't real, then you may know what reality is, and can answer the question 'Who am I?'"

This sentiment was a fitting description of the acrimony he saw around him as the band was breaking apart, and contains the wonderfully sarcastic line, "Even those tears: I Me Mine." Although Harrison can be seen playing the newly written song to Starr in the *Let It Be* film, it would not be recorded until almost a year later in 1970. The original recording of "I Me Mine" lasted only 1:30, which producer Phil Spector extended to 2:25 by repeating the first verse and chorus, and adding a cheesy orchestra for good measure.

"Dig It"	Recorded: January 24, 26, 1969	McCartney *(piano)*	Lennon *(vocals)*
(Lennon-McCartney-Starkey-Harrison)	UK Release: May 8, 1970	Harrison *(lead guitar)*	Starr *(drums)*

Additional contributors: Billy Preston (organ)

More of a jam session than a composition, this fifty-second song was only the second song credited to all four Beatles as songwriters. Led by Lennon, the recorded version was edited down from a twelve-minute twelve-bar blues, name checking B.B. King and Doris Day along the way and referencing Bob Dylan's "Like a Rolling Stone." The song ends with spoken words from Lennon, "That was 'Can You Dig It?' by Georgie Wood. And now we'd like to do 'Hark The Angels Come,'" segueing nicely into the title track of "Let It Be."

"Let It Be"	Recorded: January 25, 26, 31; April 30, 1969; January 4, 1970	McCartney *(vocals, backing vocals, piano, maracas)*	Lennon *(bass guitar)*
(Lennon-McCartney)	UK Release: March 6, 1970	Harrison *(backing vocals, lead guitar)*	Starr *(drums)*

Additional contributors: Billy Preston (organ and electric piano) and various other musicians

In his account of "Let It Be" in *The Lyrics*, McCartney is reminded of when he was studying English literature at the Liverpool Institute High School for Boys, he'd read *Hamlet* and that the play included the couplet, "O, I could tell you, but let it be. Horatio, I am dead."

Many years later, when the band were "going through times of trouble," McCartney explains that he had a dream in which his mother came to him saying, "It'll be all right," essentially, "Let it be." (Incidentally, and also in *The Lyrics*, McCartney confessed that "still to this day I have dreams about John and George and talk to them.")

The finished track has an angelic air and easily understood lyric (taken by many to refer to the Virgin Mary), and so the song became hugely popular. Lennon wasn't so keen on the Catholic undertones and appears to have been the agency behind the sequencing of the album, where "Let It Be" is placed between Lennon as a small boy piping "Now we'd like to do, Hark the Angels Come" and "Maggie Mae," a song about a Liverpool Lime Street whore.

Like most of the early 1969 material, "Let It Be" was left on the shelf for the remainder of the year until, at the start of 1970, it was further furnished with Harrison's Leslie-toned guitar solo with added fuzz-tone, some harmonized backing vocals by Harrison and McCartney, and a George Martin score for brass and cellos.

"Maggie Mae"	Recorded: January 24, 1969	McCartney *(vocals, acoustic guitar)*	Lennon *(vocals, acoustic guitar)*
(Trad. Arr. Lennon-McCartney-Harrison-Starkey)	UK Release: May 8, 1970	Harrison *(lead guitar)*	Starr *(drums)*

A traditional Liverpool folk song dating back to the nineteenth century, "Maggie Mae" tells the tale of a prostitute who robs a "homeward bounder," a sailor returning home after a trip at sea. The song was very popular among skiffle bands in the late 1950s, and it had been a regular part of The Quarrymen's repertoire.

The song seemingly had some kind of importance for Lennon, as he was still making home recordings of the song shortly before his death in 1980. At a mere forty seconds, it is the second-shortest song released on a Beatles album.

"I've Got A Feeling"	Recorded: January 22, 24, 27, 28, 30; February 5, 1969	McCartney *(vocals, bass guitar)*	Lennon *(vocals, rhythm guitar)*
(Lennon-McCartney)	UK Release: May 8, 1970	Harrison *(backing vocals, lead guitar)*	Starr *(drums)*

Additional contributors: Billy Preston (electric piano)

The first song Lennon and McCartney had truly collaborated on since "A Day In The Life," "I've Got A Feeling" is a fusion of two partly finished songs. While McCartney's part reflects his newfound domesticity with fiancée Linda Eastman, Lennon's is more of a diary entry for the ups and downs of the previous year.

As McCartney explains in *The Lyrics*, "This song is a shotgun wedding between my own 'I've Got A Feeling' and a piece John had written, called 'Everyone Had A Bad [sic] Year.' It had been a rough year or two for John. The breakup of his marriage. His estrangement from Julian. A problem with heroin. And there was the generally poor state of affairs in the band by this time. That's encapsulated in the combination of the phrases 'everybody pulled their socks up' and 'everybody put their foot down.' Those lines refer in some way to the state of the nation, or the state of The Beatles."

Recorded during the Apple rooftop performance on January 30, 1969, the band gave a soulful delivery aided by the presence of Billy Preston on electric piano. During the production of Peter Jackson's *Get Back* documentary in 2022, artificial intelligence was used to isolate Lennon's vocals for this song. This enabled McCartney to perform the song live as a virtual duet with Lennon on his 2022 *Got Back* tour. Once more, The Beatles lead the way on the innovative use of technology in music.

"One After 909"	Recorded: January 28, 29, 30, 1969	McCartney *(vocals, bass guitar)*	Lennon *(vocals, lead guitar)*
(Lennon-McCartney)	UK Release: May 8, 1970	Harrison *(rhythm guitar)*	Starr *(drums)*

Additional contributors: Billy Preston (electric piano)

One of their earliest songs, written by Lennon back in the late '50s, "One After 909" was a staple of The Quarrymen and early Beatles repertoire. It was recorded first at the sessions for "From Me To You" in March 1963, with all but one of the four takes breaking down mid-song.

Full of imagery of American railroads and freight trains in imitation of their American influences, the band revived the song for the *Let It Be* album.

Let It Be (1970)

"It was a number we didn't used to do much," McCartney said. "But it was one that we always liked doing, and we rediscovered it. It's not a great song but it's a great favorite of mine."

Recorded live during the famous rooftop performance at Apple, it is played with *joie de vivre* and effortless musicianship, especially Harrison who delivers a guitar solo *par excellence*.

"The Long And Winding Road"	Recorded: January 26, 31, 1969; April 4, 1970	McCartney *(vocals, piano)*	Lennon *(bass guitar)*
(Lennon-McCartney)	UK Release: May 8, 1970		Starr *(drums)*

Additional contributors: Various other musicians

"One of the fascinating aspects of this song is that it seems to resonate in very powerful ways," McCartney says of this famous track in *The Lyrics*. "For those who were there at the time, there seems to be a double association of terrific sadness and also a sense of hope, particularly in the assertion that the road that 'leads to your door/Will never disappear' . . . Often when I write a song, I do a bit of a disappearing trick myself . . . in this case Ray Charles. . . . This is a strategy for keeping things fresh. . . . It frees you up. You discover as you get through it that it wasn't a Ray Charles song anyway; it was yours. The song takes on its own character. The road leads . . . to somewhere you never expected."

What's remarkable about his account is the total lack of reference to the controversy surrounding its recording. The basic backing track, written on the same day as the song "Let It Be" in January 1969, was essentially appropriated by Lennon, without McCartney's knowledge, and given to American record producer Phil Spector to salvage.

The result was disastrous. Smothered in tasteless orchestration, Spector's solution was alien to the back-to-basics idea behind the *Get Back* project. (Admittedly, the orchestration was used in part to veil

Lennon's atrocious bass playing on the backing track, which is so bad you can almost "hear" McCartney "grin" at such incompetence at 1:59.) When McCartney *did* eventually hear Spector's patch-up job on his song, he was understandably incensed, and tried in vain to stop it.

Then, once he'd made sure his solo album, *McCartney*, would be released ahead of *Let It Be*, he swiftly announced he'd left the band. This rather checkered backstory is a shame for one of the most beautiful songs, musically and lyrically, that McCartney ever wrote, and that's saying something. In *McCartney 3, 2, 1*, paraphrasing Mozart, McCartney tells Rick Rubin that he "write[s] the notes that like each other," as if that were the easiest thing in the world.

"For You Blue"	Recorded: January 25, 1969	McCartney *(piano)*	Lennon *(slide guitar)*
(Harrison)	UK Release: May 8 1970	Harrison *(vocals, acoustic guitar)*	Starr *(drums)*

This straightforward twelve-bar blues song was written by Harrison for his then-wife Pattie Harrison. Lennon contributes lap steel slide guitar to the track, while Harrison namechecks Elmore James, the Delta blues slide guitar player. The track does not have a bass part.

"Get Back"	Recorded: January 23, 27, 28, 30; February 5, 1969	McCartney *(vocals, bass guitar)*	Lennon *(harmony vocals, lead guitar)*
(Lennon-McCartney)	UK Release: April 11, 1969	Harrison *(rhythm guitar)*	Starr *(drums)*

Additional contributors: Billy Preston (electric piano)

The band's nineteenth British single, "Get Back" was the first release from their 1969 "back-to-basics" project. However, as McCartney points out in *The Lyrics*, for him, "Get Back" developed another meaning, "We were a damn good little band. The four of us just knew how to fall in with each other and play, and that was our real strength.

That made it all the more sorrowful to think that our breaking up was almost inevitable. So there's a wistful aspect to 'Get Back.' The idea that you should get back to your roots, that The Beatles should get back to how we were in Liverpool. And the roots are embodied in the style of the song, which is straight-up rock and roll. Because that was definitely what I thought we should do when we broke up—that we should 'get back to where we once belonged' and become a little band again."

Curiously, the song started life as a satire on the controversy about the UK's contemporary immigration policy. Conservative politician Enoch Powell had made his racist "rivers of blood" speech in April of 1968 in which he'd claimed immigration into the UK would cause a future race war. And so, McCartney's first version of "Get Back" had been associated with the impending flight of Kenyan Asian refugees to Britain, with an original lyric about a Pakistani "living in another land," with the lines "Don't dig no Pakistani taking all the people's jobs/Get back to where you once belonged." McCartney had meant the song as a satire on racism, but the ethnic situation in the UK was so volatile after Powell's prediction that McCartney thought better of it.

AFTERWORD

"The Army That Never Was"

The Beatles continue to grab the headlines even today. In December 2023, as another anniversary of John Lennon's murder approached, *The Guardian* published an article which claimed the band endured because they "transcend time, geography, demographics, and personal taste," and that The Beatles are "like a snapshot of a particular period of time that never fades." Albums like *A Hard Day's Night* "almost six decades after its release, sound even more exuberant, more joyful—more perfect—than ever."

As mentioned on page 192, Paul McCartney's "Blackbird," his tribute to the civil rights movement, achieved a renewed global fame by being covered by Beyoncé on her *Cowboy Carter* album, released on March 29, 2024. Such enduring fame is without doubt one of the reasons why, in mid-May of 2024, the BBC reported that, according to the London Sunday Times Rich List, "McCartney has become the first UK musician to become a billionaire. . . . The former Beatle . . . boosted his wealth by £50m in the past year with touring, the lucrative value of his back catalogue and Beyoncé's cover of the classic track he wrote in 1968, 'Blackbird,' helping him achieve the status."

In April of 2024, the London *Evening Standard* had claimed that The Beatles song "Now And Then" was the previous year's "most significant release, both a time capsule of our greatest ever band, and something genuinely new." The band were not only "extraordinarily brilliant at what they did" but also represented "the great counter-narrative of post-war Britain, a journey that links austerity and the end of rationing with IG and TikTok; that travels from the sexual revolution of the sixties to the creative insurgence of AI."

John Lennon also hit the BBC headlines in late May of 2024 when his Framus twelve-string Hootenanny acoustic guitar, used to play "You've Got To Hide Your Love Away" in the 1965 movie *Help!* sold

Afterword

for $2.9 million via a telephone bid at the Hard Rock Cafe in New York, becoming the most expensive Beatles instrument ever sold at auction. David Goodman, chief executive of Julien's Auctions, said, "This guitar is not only a piece of music history but a symbol of John Lennon's enduring legacy. Today's unprecedented sale is a testament to the timeless appeal and reverence of The Beatles' music and John Lennon."

2024 also saw the announcement that Oscar-winning film director Sir Sam Mendes is to make four separate movies about The Beatles—one from each band member's perspective. According to the BBC: "The films, made by Sony Pictures and Sir Sam's Neal Street Productions, will be released in cinemas in 2027." Producer, Dame Pippa Harris, was quoted by the BBC as saying, "We intend this to be a uniquely thrilling, and epic cinematic experience. . . . To have The Beatles' and Apple Corps' blessing to do this is an immense privilege."

This book has paid tribute to history's greatest architects in the writing, recording, and artistic presentation of popular music. On a song-by-song basis, we have tried to show that, even though The Beatles's attitude to musical composition was intuitive rather than that of formally trained musicians, the science of their tradecraft was in the way that they and George Martin, the "fifth Beatle," creatively used recording technology to realize their musical ambitions. The sheer brilliance of these creative solutions to musical composition is perhaps the main reason for their continued longevity. The band's attitude toward recording was "'Try it. Just try it for us,'" according to Mark Lewisohn quoting McCartney in the book *The Beatles: All These Years*, "'If it sounds crappy, OK, we'll lose it. But it might just sound good.' We were always pushing ahead: Louder, further, longer, more, different."

In November of 2024, Disney+ dropped *Beatles '64*, a fascinating documentary movie produced by Martin Scorsese and directed by David Tedeschi. Featuring never-before-seen electrifying footage of McCartney, Lennon, Harrison, and Starr and legions of young fans during the height of Beatlemania, the documentary is an exhilarating filmic record of the band's arrival in New York City in February 1964 and their legendary live appearance on the *Ed Sullivan Show*, which solidified their status as the biggest band in the world. The band emanate a seemingly limitless, almost otherworldly energy, laughing and goofing about, as Lennon comments at one point, "The Beatles and their ilk were created by the vacuum of non-conscription . . . we were the army that never was. We were the generation that were allowed to live."

INDEX

A

Abbey Road, 219–234
Abbey Road (studio), 7–8, 135
"Across the Universe," 237–238
"Act Naturally," 91
ADT (artificial double tracking), 165
"Albatross" (Fleetwood Mac), 230
Alexander, Arthur, 18–19, 41
"All I've Got To Do," 41–42
"All My Loving," 42–43
"All Together Now," 215
All We Are Saying (Sheff), 102, 126, 129
All You Need Is Ears (Martin), 55, 97, 159, 219
"All You Need Is Love," 103, 175–176
Alstrand, Dennis, 42
Amos, Tori, 189
"And I Love Her," 64–65
"And Your Bird Can Sing," 124
animation, 213–217
"Anna (Go to Him)," 18–19
"Another Girl," 89–90
antiphony, 46–47, 50
"Any Time At All," 68

Asher, Jane, 40, 64, 89, 197–198
"Ask Me Why," 21
Aspinall, Neil, 120, 152

B

"Baby It's You," 10–11, 28–29
"Baby's In Black," 76–77
"Baby You're A Rich Man," 174–175
"Back in the USSR," 16, 180–181, 225
Badman, Keith, 9
Barrett, Richie, 13
BBC Radiophonic Workshop, 95
Beach Boys, The, 91, 118, 180
Beatlemania, 3, 35–53, 58–59, 247
Beatles, Forever, The (Schaffner), 58
Beatles, The (White Album), 177–211
Beatles, The: All These Years (Lewisohn), 18
Beatles, The: Get Back (documentary), 16, 63, 226
Beatles, The: In Their Own Words (Miles), 105–106

Beatles, The: Off the Record (Badman), 9, 11, 42
Beatles '64 (documentary), 246–247
Beatles and their Revolutionary Bass Player, The (Alstrand), 42
Beatles For Sale, 42, 73–83
Beato, Rick, 16
"Because," 228–229
"Being For The Benefit Of Mr. Kite!," 144–146
Berio, Luciano, 155
Bernstein, Leonard, 113
Berry, Chuck, 77, 99
Beyoncé, 192–193
Beyond the Fringe (comedy revue), 4
Bicknell, Alf, 120
"Birthday," 199–200
Black, Cilla, 144
"Blackbird," 191–193, 245
Bloomfield, John, 2
"Blue Jay Way," 164–165
Bowie, David, 15–17, 47, 200
Boyd, Pattie, 89, 112, 120, 222
"Boys," 20, 29
Browne, Tara, 154
BTR (British Tape Recorder), 7–8
Buckinghams, The, 72
Burns, Robbie, 12
Byrd, Charlie, 75

C

"Can't Buy Me Love," 65–67
Can't Buy Me Love: The Beatles, Britain, and America (Gould), 88
"Carry That Weight," 232
Catsoulis, Jeannette, 35
Cavern Club, 4
"Chains," 19–20
Chappell Recording Studios, 165
Charles, Ray, 181
Civil, Alan, 125
Clapton, Eric, 5, 89, 188, 207, 227–228
Cleave, Maureen, 102, 160
Colbeck, Cameron, 28
"Come Together," 221
Complete Beatles Recording Sessions, The (Lewisohn), 40, 155, 164, 193, 199, 208
"Continuing Story of Bungalow Bill, The," 185–186
Cook, Peter, 4
Cooke, Richard A., III, 186
Cooke de Herrera, Nancy, 186
Coxon, Graham, 32
Cramer, Floyd, 19
Crosby, David, 122, 136
"Cry Baby Cry," 208–209

D

David, Mack, 28
"Day In The Life, A," 153–158
"Dear Prudence," 181–182
Decca Records, 4
Dekker, Thomas, 232
"Devil In Her Heart," 50–51
"Dig A Pony," 237
"Dig It," 239
Dixon, Luther, 28
"Dizzy Miss Lizzy," 96
"Doctor Robert," 126
Donovan, 5
"Don't Bother Me," 37–38, 43–44, 89
"Don't Pass Me By," 195–196
double-tracking, 9, 165
"Do You Want To Know A Secret," 29–30, 48
"Drive My Car," 99–100
Dylan, Bob, 62, 88, 108, 236, 239

E

Edison, Thomas, 223
Ed Sullivan Show, 57–58, 71
"Eight Days A Week," 79–80
"Eleanor Rigby," 6, 113–115
Eltham, Stuart, 193
Elton John Band, 15
Emerick, Geoff, 10–11, 36–37, 39, 111, 121, 129–131, 141, 150, 222, 227

EMI (record label), 3–4, 6, 12, 39–40, 66, 151, 196
"End, The," 233–234
Epstein, Brian, 4, 12, 19, 24, 55, 171, 179
Evans, Mal, 120–121, 155
Everett, Mark Oliver, 172–173
Everett, Walter, 60–61, 121
Everly Brothers, The, 19, 23, 80
"Everybody's Got Something To Hide Except Me And My Monkey," 202–203
"Everybody's Trying To Be My Baby," 83
"Every Little Thing," 81

F

Faithfull, Marianne, 120
Fallon, Jack, 196
Fame, George, 128
Farrow, Prudence, 181
Fitzgerald, Ella, 68
"Fixing A Hole," 141–142
Fleetwood Mac, 230
"Flying," 164
Fonda, Peter, 122
"Fool On The Hill, The," 163–164
"For No One," 125
"For You Blue," 243
four-track technology, 38–39
Freymann, Robert, 126
frontman, 20–21

G

German language, 39–40
Get Back, 220, 222, 224–225, 234, 242–243. *See also Let It Be*
"Get Back," 243–244
"Getting Better," 140–141
Getz, Stan, 75
Gibran, Kahlil, 198–199
Gibson, Brian, 208
Gibson Maestro Fuzz-Tone distortion box, 37–38
"Girl," 105–106
"Glass Onion," 182–183
"God Only Knows" (Beach Boys), 118
"God Save The Queen" (Sex Pistols), 48
Goffin, Gerry, 19
"Golden Slumbers/Carry That Weight," 232
"Good Day Sunshine," 123–124
Goodman, David, 246
"Good Morning Good Morning," 151–152
"Good Night," 210–211
Goon Show, The, 5
Gordy, Berry, 216
Gorshin, Frank, 57
"Got To Get You Into My Life," 127–128
Gould, Jonathan, 88
Grant, Keith, 175
Grohl, Dave, 118

H

Haley, Bill, 21
"Happiness Is A Warm Gun," 188–189
Hard Day's Night, A (album), 60–72, 75, 245
Hard Day's Night, A (film), 5, 55–59, 63, 69
"Hard Day's Night, A," 60–61, 109
Harris, Pippa, 246
Harrison, Carl, 83, 207–208, 216–217, 227–228
Harrison, George, 8, 17–20, 43–44, 53, 63–65, 71, 89, 100, 103, 109, 112, 114–115, 122, 127, 142, 147–148, 167, 187–188, 215, 222–223, 238
Harrison, Louise, 194
Heath, Edward, 114
Hebron, Lucy, 81
"Hello Goodbye," 170–171, 234
Help! (album), 85–96
Help! (film), 5, 86, 245–246
"Help!," 32, 47, 86–87
"Helter Skelter," 204
Here, There and Everywhere: My Life Recording the Music of the Beatles (Emerick), 36–37, 39, 129
"Here, There And Everywhere," 117–118
"Here Comes The Sun," 227–228

"Her Majesty," 233–234
Herrmann, Bernard, 115
"Hey Bulldog," 216
"Hey Jude," 16, 206
Hinduism, 147, 167, 222
"Hold Me Tight," 48–49
Holly, Buddy, 20–21, 80
"Honey Don't," 81
"Honey Pie," 207
humor, 5

I

I, Me, Mine (Harrison), 44, 127
"I Am The Walrus," 27, 166–170, 216
"I Don't Want To Spoil The Party," 82
"If I Fell," 62–63
"If I Needed Someone," 110
Ilhan, Adem, 14–15
"I'll Be Back," 72
"I'll Cry Instead," 69
"I'll Follow The Sun," 78
"I'm A Loser," 76–77
"I Me Mine," 238
"I'm Happy Just To Dance With You," 63–64
"I'm Looking Through You," 106–107
"I'm Only Sleeping," 116
"I'm So Tired," 190–191
"I Need You," 89
In His Own Write (Lennon), 85

"In My Life," 107–108
"I Saw Her Standing There," 12–15
"I Should Have Known Better," 62
"It's All Too Much," 216–217
"It's Only Love," 92
"It Won't Be Long," 38, 41–42
"I've Got A Feeling," 240–241
"I've Just Seen A Face," 93–94
"I Wanna Be Your Man," 49–50
"I Want to Hold Your Hand," 35, 38–39, 55
"I Want To Tell You," 127
"I Want You (She's So Heavy)," 226–227
"I Will," 197–198

J

Jackson, Peter, 16, 63, 226
Jagger, Mick, 5, 16, 50, 120, 124, 175
James, Dick, 214–215
Jarrat, Jeff, 226
Jarry, Alfred, 223
John, Elton, 15
Johnson, Wilko, 23
Jones, Berth, 7
Jones, Brian, 5, 117, 120
Jones, Quincy, 5, 7
"Julia," 198–199

K

"Kansas City/Hey-Hey-Hey-Hey!," 79
Kantner, Paul, 136
Kesey, Ken, 162
King, Carole, 19
King, Martin Luther, Jr., 192, 206
Klein, Allen, 229
Kramer, Billy J., 29
Kramer, Eddie, 175
Krenwinkel, Patricia, 194

L

Leary, Timothy, 126, 129, 221
Lemmy, 13–14
Lennon, Cynthia, 65, 69, 151, 203, 245–246
Lennon, John, 5, 11–12, 14–15, 26–27, 29, 32–34, 42, 45, 53, 63–64, 69, 85–87, 92, 100, 103, 105–106, 110, 122, 153–154, 160, 167, 197–199, 202, 209, 211, 221, 226, 237
Lennon, Julian, 139
Lester, Richard, 5, 59, 72, 86, 93
Let It Be, 8, 16, 235–244. *See also Get Back*
"Let It Be," 239–240
Letterman, Paul, 58
Lewisohn, Mark, 18, 155, 164, 193, 199, 208, 246
Life of Brian (film), 102–103

"Little Child," 44–45
Little Richard, 224
Little Rock Nine, 192–193
Loach, Ken, 143
Lockwood, Joseph, 136
"Long And Winding Road, The," 242–243
"Long Long Long," 206
Lothian, Andi, 57
"Lovely Rita," 48, 150–151
"Love Me Do," 1, 3, 11, 24–27
"Love You To," 113, 117
LSD, 87, 91, 98, 116, 122, 127, 129, 139–140, 142, 162, 167, 216–217, 238
"Lucy In The Sky With Diamonds," 139–140
Lyrics, The (McCartney), 24, 42, 67–68, 70, 77, 94–95, 99–100, 115, 118–119, 142, 173, 180, 184, 189, 192, 197, 199–200, 204, 229, 239, 241, 243–244

M

MacDonald, Ian, 17, 221, 229
MacNeice, Louis, 148–149
"Maggie Rae," 48, 240
Magical Mystery Tour, 4, 159–176
"Magical Mystery Tour," 162–163

Maharishi Mahesh Yogi, 163–164, 179, 181, 186, 201–203, 238
Mann, William, 38
Manson, Charles, 194, 204
Many Years From Now (Miles), 81–82, 116, 123, 126, 142, 199, 215
Marr, Johnny, 166–169
"Martha My Dear," 189–190
Martin, George, 4–7, 10–13, 17, 21, 25, 33–34, 39–40, 45–46, 55, 61, 63, 65, 67–68, 74–75, 77, 79, 89–90, 94–95, 97, 121, 125, 130–131, 135, 143–144, 149, 153, 155, 158–159, 168, 172, 174, 195, 208, 210, 219–220, 223, 228, 246
Martin, Giles, 6
Mason, Dave, 117, 174
Maxwell, James Clerk, 223
"Maxwell's Silver Hammer," 223–224
McCartney, Paul, 1, 9, 11, 14, 17, 20, 24, 27–28, 33–34, 43–45, 58, 64–65, 67–68, 70, 77, 81–82, 89, 92–95, 99–101, 104–106, 115–116, 118–119, 123, 125–126, 133, 138, 142, 148–149, 154, 173–175, 179, 184, 189–190, 192–193, 197–200, 204, 207, 211, 215, 220, 222–224, 229, 232, 234, 236–237, 239–241, 243–244

McCartney 3, 2, 1 (television), 17, 26, 43, 104, 174, 243
McGuinn, Roger, 122
McKern, Leo, 5
"Mean Mr. Mustard," 230
Mendes, Sam, 246
"Michelle," 104
Miles, Barry, 21, 65, 81, 101, 106, 199
Milligan, Spike, 5
"Misery," 10, 17, 20, 51
"Money," 52–53
Monty Python, 5, 102–103
Moog synthesizer, 220, 226, 228
Moon, Keith, 5
Moore, Alan, 150
Moore, Dudley, 4
"Mother Nature's Son," 201–202
Murphy, Cillian, 157
Mustard, John Alexander, 230

N

Nash, Graham, 5
Nichol, Jimmy, 140–141
"Night Before, The," 87–88
Nilsson, Harry, 72
"No Reply," 42, 75
Norman, Philip, 112, 191–192
"Norwegian Wood," 100–101
"Not A Second Time," 38, 42, 52
"Now and Then," 6, 245
"Nowhere Man," 101–102

O

"Ob-La-Di, Ob-La-Da," 16, 183–185
"Octopus's Garden," 225–226
"Oh! Darling," 224–225
Oldham, Andrew Loog, 49–50
"One After 909," 241–242
"Only A Northern Song," 214–215
"Only The Lonely" (Orbison), 22
Ono, Yoko, 186, 200, 202, 210, 226
Oppenheim, David, 113
Orbison, Roy, 22
Orlando, Jordan, 177
O'Shea, Tessie, 57
overdubbing, 7, 9–10, 37, 44–45, 64, 87, 90, 107, 121, 140, 152, 163, 184–185, 191, 226, 228

P

Page, Jimmy, 235
Parkes, Stanley, 86
Parlophone (record label), 4–5
Parsons, Alan, 224
Pathe Marconi Studios, 40, 67
Paul McCartney: Many Years From Now (Miles), 21, 65, 101, 133
Paxman, Jeremy, 15–16
"Penny Lane," 6, 27, 173–174
Perkins, Carl, 83
"Philadelphia Freedom" (Elton John), 15
Pickett, Wilson, 71
"Piggies," 179, 193–194
"Please Mr. Postman," 46–47
Please Please Me, 1–34, 45, 48, 52
"Please Please Me," 11, 22–24
Pollack, Alan, 21, 79, 83, 96, 102, 107, 114, 121, 125, 148, 164, 176, 206, 216–217, 233
"Polyethylene Pam/She Came In Through The Bathroom Window," 48, 231
Powell, Enoch, 244
Presley, Elvis, 110
Preston, Billy, 5, 222, 241
Proclaimers, The, 47
"P.S. I Love You," 27–28
punk movement, 47–48

Q

Quarrymen, The, 22, 241

R

Reagan, Ronald, 221
"Revolution 1," 206
"Revolution 9," 209–210
Revolution in the Head (MacDonald), 221
Revolver, 111–131
Richards, Keith, 5
Richardson, Tony, 30

Rickenbacker, George, 61
Robinson, Smokey, 41, 49
"Rock and Roll Music," 77
"Rocky Raccoon," 194–195
Rolling Stones, 37–38, 49–50, 117, 167
"Roll Over Beethoven," 48
Rowlands, Tom, 128
Rubber Soul, 97–110
Rubin, Rick, 17, 27, 43, 174, 243
"Run For Your Life," 110
Running, Jumping and Standing Still Film, The (film), 5–6

S

"Savoy Truffle," 207–208
Schaffner, Nicholas, 58
Scheff, David, 129
Scorcese, Martin, 246
Scott, Jimmy, 184
Scott, Ken, 200
Sellers, Peter, 5
Sex Pistols, 48
"Sexy Sadie," 179, 203–204
Sgt. Pepper's Lonely Hearts Club Band, 133–158
"Sgt. Pepper's Lonely Hearts Club Band," 137
"Sgt. Pepper's Lonely Hearts Club Band (Reprise)," 152
Shankar, Ravi, 147
Shapiro, Helen, 17
Sheff, David, 102, 126

"She Loves You," 37, 41, 176
"She Said She Said," 122–123
"She's Leaving Home," 143–144
Shirelles, The, 28
Shotton, Pete, 169
Shout!: The Beatles in Their Generation (Norman), 112, 191–192
Simonelli, David, 161
Smith, Alan, 33
Smith, Norman, 12, 37
"Some Other Guy" (Barrett), 13
"Something," 222–223
"Somewhere," 31
songwriters, 11
Spector, Phil, 5, 242
Starr, Ringo, 20, 82, 86, 107, 138, 195–196, 200–201
Sting, 72
"Strawberry Fields Forever," 171–173, 234
Sullivan, Ed, 57
"Sun King/Mean Mr. Mustard," 230
Sutcliffe, Paul, 26

T

"Taste of Honey, A," 9, 30–31
Tate, Sharon, 194
"Taxman," 114
Taylor, Derek, 73, 136, 165, 188, 207–208

Teenager's Turn – Here We Go (television), 21
"Tell Me What You See," 93
"Tell Me Why," 65
"There's A Place," 31–32
"Things We Said Today," 70
"Think For Yourself," 103
Thomas, Chris, 186, 199, 208
"Ticket To Ride," 90–91
"Till There Was You," 45–46
"Tomorrow Never Knows," 6, 128–131
Traffic, 117
"Twist and Shout," 33–34

V
Valli, Frankie, 21
vari-speed recording, 9–11

W
"Wait," 109
Watts, Charlie, 25
"What Goes On," 105
"What You're Doing," 82–83
"When I Get Home," 71
"When I'm Sixty-Four," 148–149
"While My Guitar Gently Weeps," 48, 187–188
White, Andy, 28
Who, The, 199–200
"Why Don't We Do It In The Road?," 196–197
"Wild Honey Pie," 185
Williams, Holly, 213
Williams, Larry, 96
Wilson, Brian, 59
Wilson, Harold, 114
Winner, Langdon, 136
"With A Little Help From My Friends," 138–139
"Within You Without You," 146–148
With The Beatles, 28, 36–53
Wonder, Stevie, 112
"Word, The," 103–104
"Words of Love," 80

Y
Yellow Submarine (film), 213–217
"Yellow Submarine," 7, 118–121
"Yer Blues," 200–201
"Yesterday," 94–96
"You Can't Do That," 71–72
"You Like Me Too Much," 92
"You Never Give Me Your Money," 229
Younger, Richard, 18
"You're Going To Lose That Girl," 90
"Your Mother Should Know," 165
"You've Got To Hide Your Love Away," 88, 245–246

"You've Really Got A Hold On
 Me," 49
"You Won't See Me," 101

Z
Zubes (lozenge), 12